AF688849

Management of Natural Disasters

WITPRESS

WIT Press publishes leading books in Science and Technology.
Visit our website for the current list of titles.
www.witpress.com

WIT*eLibrary*

Home of the Transactions of the Wessex Institute, the WIT electronic-library provides the international scientific community with immediate and permanent access to individual papers presented at WIT conferences.
Visit the WIT eLibrary at http://library.witpress.com

Management of Natural Disasters

Edited by

S. Syngellakis
Wessex Institute, UK

WIT*PRESS* Southampton, Boston

Editor:

S. Syngellakis
Wessex Institute, UK

Published by

WIT Press
Ashurst Lodge, Ashurst, Southampton, SO40 7AA, UK
Tel: 44 (0) 238 029 3223; Fax: 44 (0) 238 029 2853
E-Mail: witpress@witpress.com
http://www.witpress.com

For USA, Canada and Mexico

Computational Mechanics International Inc
25 Bridge Street, Billerica, MA 01821, USA
Tel: 978 667 5841; Fax: 978 667 7582
E-Mail: infousa@witpress.com
http://www.witpress.com

British Library Cataloguing-in-Publication Data
A Catalogue record for this book is available
from the British Library

ISBN: 978-1-78466-229-5
eISBN: 978-1-78466-230-1

Library of Congress Catalog Card Number: 2016941767

No responsibility is assumed by the Publisher, the Editors and Authors for any injury and/or damage to persons or property as a matter of products liability, negligence or otherwise, or from any use or operation of any methods, products, instructions or ideas contained in the material herein. The Publisher does not necessarily endorse the ideas held, or views expressed by the Editors or Authors of the material contained in its publications.

© WIT Press 2017

Printed by Lightning Source, UK.

All rights reserved. No part of this publication may be reproduced, stored in a retrieval system, or transmitted in any form or by any means, electronic, mechanical, photocopying, recording, or otherwise, without the prior written permission of the Publisher.

Preface

The present volume comprises a selection of articles from the Transactions of the Wessex Institute dedicated to disaster management. The focus of this collection is on addressing the challenges arising from extreme natural phenomena, with descriptions of methods for assessing their occurrence probability and of measures for mitigating their intensity and detrimental effects.

The first group of articles are concerned with general strategies for risk assessment and mitigation, some of them providing examples in the context of various kinds of natural disasters. Räikkönen *et al.* propose a decision-making tool based on assessing the economic impact of mitigation measures and costs of consequences, as well as the associated uncertainty; their methodology is demonstrated trough two case studies involving a storm and an earthquake. In their disaster severity assessment, Hasani *et al.* put emphasis on communities' differing coping capabilities and classify the possible impact scenarios; their tool design is based on data from a wide range of natural disasters. Cuadra *et al.* examine human attitudes among poor communities towards a specific mitigation measure, namely relocation, referring to cases of tsunami, earthquake and flooding. According to Audefroy, droughts, floods and hurricanes are potentially linked to climate change and predictions from his research may contribute to the design of suitable adaptation strategies. National strategies towards natural disasters are described in three contributions: Oktay outlines Turkey's general plan for a coordinated response to the consequences of earthquakes and other natural disasters; Soriano and Hoffmann provide an account of various state organisations established mainly to mitigate the consequences of disasters from geological and hydrological processes in Brazil; finally, Al Hmoudi & Aziz put emphasis on the training needs of communities so that they respond effectively to early warnings of all kinds of natural disasters in United Arab Emirates.

The second part of the book is concerned with disasters from specific natural causes starting with a group of ten articles on floods. Escalante-Sandoval and Garcia-Espinoza offer a detailed probabilistic assessment of flood frequency in Mexico; Solín *et al.* combine flood hazard with vulnerability and resilience of municipalities in their multi-criteria decision analysis of flood risk in Slovakia; Coates *et al.* apply agent-based modelling and simulation to assess how the response of small and medium enterprises to flood events affects their business continuity in the United Kingdom; Huttenlau *et al.* assess flood risk in terms of potential financial losses in an Austrian study area; Moccia and Sgobbo propose interventions enabling the participation of private investment to measures that aim to reduce the impact of extreme rainfall in the periphery of Naples metropolitan area; Miguez *et al.* used the concept of "flood risk index" to assess design solutions for mitigating flood hazard in a Brazilian urban environment; Peddemors and Willemse focus on the role of "sleeper" dykes in reducing the impact of flooding in the Dutch province of Zeeland and the possibility of making them more effective through changing geography by design; Takezawa *et al.* used a questionnaire and discussions to assess the flood preparedness of hospitals and clinics in two major Japanese cities; Budhakooncharoen and Dhabhisara focus on highway flooding in Thailand and identify the relative importance of highway, hydrological and topographical characteristics as possible causes of it using a multi-criteria analysis; Awopetu *et al.* used questionnaires to gauge the impact of flooding on the agriculture, education, health, housing and sanitation in affected communities of Nigeria.

The subsequent four articles are concerned with earthquake-related hazards: Menteşe *et al.* test the ability of an analysis tool to provide an integrated seismic risk, coping capacity and risk management output for the Istanbul Metropolitan area; Morán-Rodríguez and Novelo-Casanova apply a methodology involving structural, non-structural, functional and social-administrative elements as well as weighting according to the importance of each element to assess the seismic vulnerability of Mexican health facilities; García-Ayllón Veintimilla *et al.* apply new techniques for urban aftershock management and damage assessment based on microzoning and a territorial plan; Gotoh and Takezawa identify factors influencing inundation and use neural network modelling to estimate potential damage caused by tsunamis.

A final group of three articles address issues related to landslides: Abdullah describes action plans for slope management in Malaysia as a means of reducing risk and losses due to landslides; Oh *et al.* apply a method they developed for early landslide warning based on real time rainfall data; Fabbri and Chung predict landslide hazard and assess the associated risk using the mathematical framework of favourability function modelling and a software for spatial target mapping.

The volume provides a unique global perspective and focus on recent developments over a wide range of topics that cannot be found compiled in similar, currently available publications in this field. As such, it is expected to be a welcome addition to the relevant literature among researchers and engineers working on risk assessment and mitigation of intensity and consequences of natural disasters in academic and research environments as well as governmental or professional, national and international organisations.

Stavros Syngellakis
The New Forest, 2017

Acronyms

ABM	Agent Based Models
AHP	Analytic Hierarchy Process
ARIO	Adaptive Regional Input-Output
CBA	Cost-Benefit Analysis
CGI	Common Gateway Interface
CRC	Czech Red Cross
CVAT	Community Vulnerability Assessment Tool
DEM	Digital Elevation Model
DRI	Disaster Risk Indicator
DRM	Disaster Risk Management
DSA	Disaster Severity Assessment
DTM	Digital Terrain Model
EMI	Earthquake and Megacities Initiative
ES	Emergency Situations
EWS	Early Warning System
FRI	Flood Risk Index
GFCS	Global Framework for Climate Services
GIS	Geographic Information System
GUI	Graphical User Interface
HDI	Human Development Index
HRM	High Resolution Model
IDB	Inter-American Development Bank
IFRC	International Federation of the Red Cross and Red Crescent
IMM	Istanbul Metropolitan Municipality
IPCC	Intergovernmental Panel on Climate Change
ISDR	International Strategy for Disaster Reduction
LCC	Life Cycle Costing
LDW	Logical Decisions for Windows
LiDAR	Light Detection And Ranging
LW	Landslide Warning
LWS	Landslide Warning System

MADM	Multi-Attribute Decision-Making
MCA	Multi Criteria Analysis
MCDA	Multi Criteria Decision Analysis
MLS	Multi-Level Safety
MODM	Multi-Objective Decision-Making
MSS	Mobile Switching Station
NGO	Non-Governmental Organisation
OGC	Open Geospatial Consortium
OS	Ordnance Survey
RMI	Risk Management Indicator
RP	Return Period
SME	Small and Medium Enterprises
SPC	State of Public Calamity
VGE	Virtual Geographical Environment
WMO	World Meteorological Organization
WPS	Web Processing Service
WSGI	Web Service Gateway Interface

Contents

Assessing the economic impacts of crises: a decision-support approach to long-term strategic planning

M. Räikkönen[1], K. Pilli-Sihvola[2], S. Kunttu[1], J. Yliaho[1], M. Jähi[1], G. Zuccaro[3] & D. Del Cogliano[4]
[1]*VTT Technical Research Centre of Finland, Finland*
[2]*Finnish Meteorological Institute, Finland*
[3]*Università di Napoli Federico II, PLINIVS-LUPT Study Center, Italy*
[4]*Eureka Comunicazione Telematica, Rome, Italy*

Abstract

The assessment of economic impacts of crises and risk reduction generated by different mitigation measures is a multidimensional problem and often a challenging task. A major issue hindering the process is a lack of structured information. In particular, intangible losses are difficult to evaluate monetarily. Furthermore, the uncertainty related to the operational environment is difficult to estimate. This paper proposes a practical economic approach that aims at assessing different measures to mitigate the impacts of the crises. The purpose is to contribute to the understanding of the decision-making related to crisis management so that the goals of the decisions can be reached. The approach is mainly aimed at enhancing decision-making on a strategic level and at supporting crisis-management training. Additionally, a software tool based on the approach will be developed to make the assessment procedure usable in practice. The approach is demonstrated in two pilot cases: extreme winter storm in North of Finland and a major earthquake in Italy. The paper is based on research carried out in the CRISMA integration project which is funded by the European Community's Seventh Framework Programme.
Keywords: cost, benefit, crisis, decision-making, mitigation measure, economic impact.

1 Introduction

Nowadays, the risks of crises are surrounding us practically everywhere. Although partial risk reduction (or mitigation) is feasible by introducing and implementing different measures to mitigate the impacts of the crises, they continue to occur and can cause severe damage to physical assets, lives and livelihoods [1]. Hence, decision-makers at all decision-making levels are pressured to find ways to cope with the impending crises, increase awareness of the overall impacts and to have a thorough understanding of the economic impacts of the crises. It is evident that decision-makers usually find themselves confronted with the situation where they have to balance the uncertainties in potential impacts, economic costs, time, resources, the level of information and other possible priorities [2]. In this respect, estimates of and other information on the costs and benefits of the mitigation measures and crisis impacts are crucial for decision-making and for the development of strategies and measures to prevent or reduce the impacts.

To support reliable, cost-effective, efficient, and transparent decision-making on the mitigation measures, different investment appraisal and cost assessment methods can be applied. A considerable amount of research is done in this area and the general features of these methods are well known [3–6]. Additionally, a variety of methodological approaches on the economic assessment of alternative risk mitigation measures exists. These approaches are based, for example, on cost-benefit analysis (CBA) and life cycle costing (LCC) [3, 7–9].

It is generally accepted that the assessment of the mitigation measures should cover both the investment and operating costs and the benefits (i.e. the avoidance or reduction of negative consequences of the crises) of a measure or a set of measures. Mitigation investments include, for example, physical assets, engineering techniques and hazard-resistant constructions, as well as improved policies, plans and actions taken to avoid, reduce the severity of, or eliminate the adverse impacts of crises. Furthermore, costs of rescue and emergency services in the response phase should be included [3]. Often the assessment is a challenging task, as the methodology to assess economic losses is not standardized [10, 11].

The costs and benefits of the mitigation measures can be further divided into direct and indirect as well as tangible and intangible. However, often the assessment of risk mitigation measures focuses almost exclusively on the direct costs, rather than indirect or intangible costs. The main reason is that the direct costs are most straightforwardly expressed in monetary terms [3, 10, 12–14]. It is also relatively simple to find data on direct costs. This indicates that the assessments are often under pressure to demonstrate short-term effects [3, 10, 12–14]. Additionally, the assessments tend to concentrate only on direct investments on "hard" mitigation measures, such as infrastructure investments, rather than on non-structural measures, such as small-scale mitigation actions, monitoring and warning systems as well as emergency response capacities [3].

An economic assessment usually entails considerable uncertainties. For example, data sources can be insufficient or highly aggregated [3] and uncertainties related to the operational environment create confusion. Uncertainties can enter the assessment at different points (e.g. data uncertainty, parameter uncertainty and model uncertainty) which all affect the decision-makers' perception of the usefulness of the results [2]. However, only a few practical analyses to integrate uncertainty aspects systematically into the decision-making exist and, as a result, almost no practical solutions appear to be available. Furthermore, it is here argued that many analytical and normative models to aid decision-making on mitigation measures are often too theoretic and complicated to be used in a practical context (see e.g. [3, 10, 13–16]).

2 Objectives and method

The aim of this paper is to contribute to the practical assessment on the economics – costs and benefits – of mitigation measures (and assets, more broadly) for crisis management in a way that the goals of the decisions can be reached. This paper is based on research carried out within the CRISMA integration project, funded from the European Community's Seventh Framework Programme. CRISMA aims to support both public and private crisis managers and decision-makers in short- and long-term planning, training and reviewing of crisis preparedness as well as response organisations, infrastructure and personnel [1]. As a main outcome, CRISMA will develop a simulation-based decision support system. One component of the CRISMA decision support system is a software tool based on the economic evaluation approach presented in this paper.

The underlying research question of this paper is: *how to assess in a practical way the investments aiming at reducing the negative impacts of crises?* The two more specific objectives of the paper are:

- *To propose a practical framework for the economic assessment of mitigation measures that will serve especially the strategic decision-making.*
- *To examine how to manage uncertainty in the assessment of mitigation measures.*

The main research methodology is constructive research. A content analysis was used to examine and compare the past and present methods of investment appraisal, CBA and LCC analyses and to discuss the different aspects of decision-making and uncertainty management in this context. The actual approach and software prototype development is based on problem solving and solution building. The approach is demonstrated in two pilot cases: extreme winter storm in the North of Finland and major earthquake in Italy.

3 Approach

3.1 Assessment of costs and benefits of different mitigation measures

The primary purpose behind the assessment of the costs and benefits of different mitigation measures is to determine the long-term implications of decisions. An important aspect is to assess an optimal level for the risk mitigation investment [17]. The assessment procedure for risk mitigation investments can be more or less analytical in depth, depending on the context and the actual case within that context (e.g. coastal submersion, earthquake, extreme winter storm, chemical spill and a large traffic accident in CRISMA). The level and type (qualitative, quantitative, or semi-quantitative) of detail in the assessment should be consistent with the level of the decision (e.g. local, national, or EU) [2].

3.2 Proposed assessment procedure

The approach proposed in this paper is based on the assessment of avoided or reduced negative consequences (benefits) generated by different mitigation measures. It can be considered as a decision-support approach to be applied in the investment planning phase. The main target group of the approach is authorities on different levels of public decision-making. In addition, the economic assessment can produce information for other purposes, for instance for insurance companies, private sector investors or international aid providers. The approach supports the decision-makers as it creates a common understanding of the decision alternatives and their possible consequences before the decision takes place. To meet the expert competence requirements, the analysis should be carried out in an expert session. One key for making successful decisions is the ability of the decision-maker to recognise and use external and internal sources of economic and damage information (e.g. databases, statistics, expert judgement) and to read weak signals. Furthermore, the decision-maker needs to efficiently take advantage of existing information and knowledge.

The proposed approach follows the passage from vulnerability analysis and potential damage estimation to loss assessment, focusing on answering the question of how such “damage” may be converted into economic losses. The basis for the assessment is a baseline scenario which presents the impacts of the crises without any implemented mitigation measures. This baseline scenario is then compared with a scenario including some mitigation measures(s) to evaluate the benefits, i.e. change (reduction) in damage and cost. The procedure (see fig. 1) can be applied to determine the economic effectiveness of different mitigation investments and to compare them with each other.

The approach aspires to systematically consider all present and future costs and benefits of mitigation investments. The ‘benefit’ part usually includes intangible benefits (e.g. lives saved, personal memorabilia not destroyed), which are difficult to measure solely in economic terms. However, the assessment should include each cost and benefit that is appropriate with each specific measure, tailored to fit the situation. The assessment approach comprises seven general, pre-defined cost and benefit types: investment and operating costs of mitigation measures, rescue and emergency costs in the response phase, costs/benefits of the impacts on human health, property, lifelines, nature and agriculture.

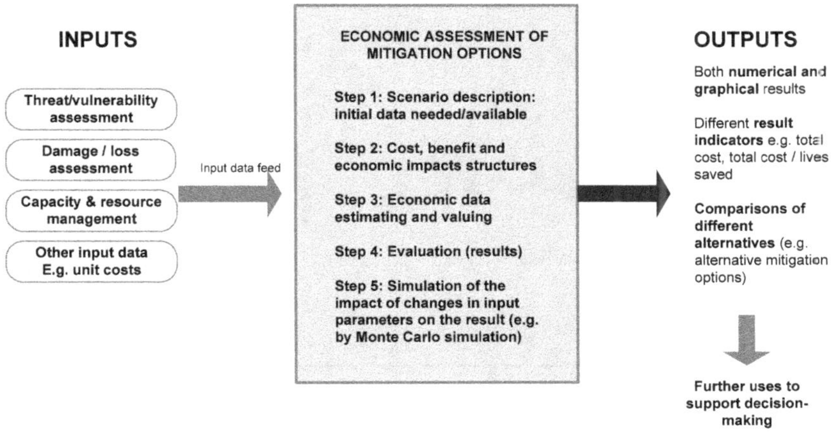

Figure 1: The process for the economic assessment of mitigation options.

There are a number of alternative methods for valuing the impacts of crises (e.g. shadow prices, replacement cost method, production method, substitute or proxy method, change in earnings, hedonic pricing, travel cost method, willingness to pay) which are suitable for different purposes, and vary in how accurately they present the value of damages. The chosen method should always reflect the decision situation at hand, the possibility to assign monetary values for different costs and benefits and data availability.

3.3 Results of the proposed approach

As a result of the assessment, different types of result indicators, e.g cost summaries as well as graphs, can be illustrated. Effectiveness and efficiency are key outcomes in determining the best option from a set of measures. Typically, cost-effectiveness relates the cost of a mitigation measure to its key outcomes or benefits (see for example [18]). The number of lives saved is an obvious unit of effectiveness. Opportunity to compare the benefits of different physical investments, plans and actions enhances the transparency of decision-making and also supports the long-term planning and decision-making. An example of a result graph is presented in fig. 2.

In all, the economic assessment of the mitigation measures has to be flexible to handle different crises with distinct cost factors. The level of available data varies case by case and the assessments have to allow calculations based on detailed data as well as rough calculations with imprecise data.

3.3.1 Reliability of the results

The key issue in considering the usefulness of the results of the economic assessment for decision-making is the reliability of the assessment, which is, in this case, mainly dependent on the completeness of cost and benefit structures and the availability of monetary values for parameters.

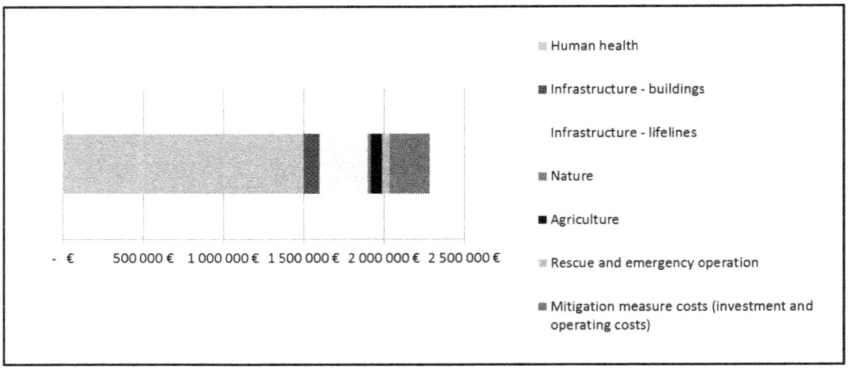

Figure 2: Example of a result graph generated by the tool based on the developed approach.

The cost and benefit structures are always case specific and need to be defined by a user of the final software tool. Thus, inclusion of all relevant cost and benefit categories and parameters is dependent on the user's expertise. To assist the user in the cost and benefit structure definition, a pre-defined structure is given as a starting point. However, the default structure might require substantial modification before the cost and benefit parameters are appropriate for the case.

The economic assessment requires a considerable amount of data which cannot be verified because they do not necessarily exist at the time decisions are made. Thus, the best available estimates of monetary values need to be defined. For example, if market values for costs and benefits are available, they can be used. If not, different valuation methods, mentioned in section 3.2, can be applied. However, there always needs to be a balance between time, resources, and the level of information: typically, decisions have to be made with less than perfect data and in a given timeframe. One method to assess the effect of uncertain monetary values on result indicator values is sensitivity analysis, presented briefly in the next section.

3.4 Sensitivity analysis

The economic assessment of mitigation measures is mostly based on the estimates of inherently uncertain future values of the economic parameters. The robustness of the assessment for a change in the parameters can be evaluated by a sensitivity analysis. The simplest way to conduct a sensitivity analysis is to change an initial value of the cost or benefit parameter and re-calculate the result indicator values, i.e. to conduct a what-if analysis.

In the proposed approach, the sensitivity analysis is performed with Monte Carlo simulation using statistical triangular distribution. Triangular distribution is defined by three values: expected minimum, maximum and mode, which is the most probable value. This distribution is selected because it is simple to understand and to define by an average user with limited knowledge about statistical distributions, and because in practice much more than minimum, maximum and mode estimates are not available for the distribution definition.

The sensitivity analysis can be performed for one or several different cost and benefit parameters at a time, and the result is the variation in the result indicator values. By selecting only one or few cost and benefit parameter values to be simulated, it is possible to study the effect of certain parameters on the result indicator values.

4 Case studies

4.1 Case study I

The first case study describes a crisis situation due to extreme winter conditions and a heavy winter storm in the North of Finland, leading to substantial damage to the electricity distribution network due to the snow load and broken trees. After the storm has passed the area, a prolonged cold-spell arrives. Due to the difficult conditions, snow clearing and repairing the damaged networks are put under heavy pressure. This combined with long power outages causes a need for evacuation of the vulnerable population.

4.1.1 Mitigation measure investments (investment costs, operating costs)

The most relevant measure to reduce the impacts of the crisis is the training of the emergency and rescue personnel and improved preparedness of the stakeholders involved, as they have the potential to decrease the adverse impacts substantially. These with the potential impacts will be applied in the final tool to assess their cost-effectiveness. Moreover, an important mitigation measure in extreme weather events is a multi-hazard early warning system which will also be evaluated from a financial perceptive.

Regulations on fire places and generators would pose a cost burden on the households and institutions, such as elderly people's homes and hotels accommodating vulnerable population. The benefits of the regulation will be evaluated with the tool.

The most significant mitigation measure would be major investments in the electricity network and housing – ground electricity cables, fire places and generators – to reduce the vulnerability of the electricity distribution network and the affected people and property. The investment cost of the ground cables is high and falls to private electricity distribution companies, but they have the potential to reduce the impacts of the event practically to zero, as snow and fallen trees cannot damage the distribution network. However, as the main users of the tool will be officials, this measure will not be evaluated.

4.1.2 Rescue and emergency costs

The costs of rescue and emergency operations are the main focus of Case Study I. The main cost category among the rescue and emergency costs is the compensation for the overtime hours for each member of the permanent staff. Furthermore, the cost of the working hours of any back-up staff is calculated.

The cost of use of helicopters and snowmobiles is calculated. Daily cost consists of number of rescue units × the cost of unit × the use of unit (hour or km). The total rescue and emergency costs are the daily cost times the number of days the event lasts.

4.1.3 Direct damage costs

The main focus of the direct impact calculation is on the impacts on human health: injured and evacuated people, and possible fatalities, and the impacts on the maintenance of the electricity and road networks. For road network, the costs occur due to the increased number of accidents and clearing the roads from snow. Quick repairing of the electricity network is crucial for the impact mitigation and to avoid evacuations. In challenging conditions, this requires increased number of workforce and equipment. The cost and benefit of overtime and back-up staff will be evaluated with the tool. An important cost category is the compensation for cross-border help, in case additional helicopters are needed or patients have to be sent to hospitals outside Finland.

4.1.4 Indirect damage costs

The most severe indirect, tangible cost category is the increase in the price of electricity network maintenance and damage repair of houses, if there is demand surge in the area and workforce arrives from other areas of Finland or from neighbouring countries. Indirect, intangible damage in the winter storm event occurs if there are long-term impacts on human health.

4.2 Case study II

The second case study describes a crisis situation due to an earthquake in Italy (Central Eastern Area), followed by a fire, caused by a cascading effect of the earthquake (an electrical short circuit generates a fire that spreads over a wide part of the area struck by earthquake).

4.2.1 Mitigation measure investments (investment costs, operating costs)

The most important measure to reduce the costs resulting from the effects of the earthquake is a decrease in vulnerability of buildings with interventions to improve their seismic performance (move the vulnerability class), at least for the building located close to the life-lines.

The cost of "Mitigation Policies" is related to the activities provided for the risk reduction. The risk can be mitigated by working on the consistency of the value exposed to the hazard and on its vulnerability for the different kinds of buildings.

Since the mitigation activities are focused on vulnerability reduction, for the buildings on which mitigation is done there is an inverse relation between mitigation costs and building reconstruction/rehabilitation costs [19].

This category included the costs related to the activities provided for the communication of the emergency plan and for the information/dissemination. For the fire hazard, the mitigation activities are mainly represented by fire spreading barriers.

This category also includes the costs related to the activities provided for the communication of the emergency plan and for the information dissemination.

4.2.2 Rescue and emergency costs

The main costs under rescue and emergency category are [19]:

- *Evacuation direct costs:* related to the activities provided for the evacuation of the population from the affected area, according to the specific needs included in the individual municipal plans.
- *Evacuation assistance costs:* related to the activities provided for giving continuous assistance to the population evacuated from the emergency area.
- *Emergency costs:* related to the activities and to the units engaged in the emergency phase as operating structures, vehicles, equipment, and volunteers. For example, the cost related to the number of people committed in each operating structure to manage the emergency (Civil Protection Department personnel, police, forest corps, etc.) and the organization of equipped area are included in this group.

4.2.3 Direct damage costs

The main direct damage costs in Case II are the following [19]:

- *Rubble cleaning-up costs:* related to the activities provided for removal and disposal of rubble from road.
- *Reconstruction costs:* related to the activities provided for removing the physical damage to capital assets including buildings, infrastructure and industrial plants through "in place" or "delocalized" reconstruction.
- *Rehabilitation costs:* related to the activities provided for removing the physical damage to capital assets, including buildings, infrastructure and industrial plants by the rehabilitation of damaged.
- *Delocalization costs:* related to the economic incentives provided for encouraging the consensual delocalization of part of the population and of the economic functions not compatible with earthquake risk proneness. The percentage of resident population and of the number of economic factors which are delocalized is linked to the percentage of the volume of residential buildings and of industrial plants which will be subject to delocalized reconstruction interventions.
- *Human health intervention costs:* related to the health care management as the implementation of advanced medical structures, the strengthening of the existing local health structures, the identification of poor people (elderly and disabled), the psychological and social assistance.
- *Back home costs:* related to the activities provided for the "back home" of the population evacuated from the emergency area.

4.2.4 Indirect damage costs

The most important indirect, tangible cost categories are [19]:

- *Decrease in local value added due to psychological effects:* estimated through the comparison of the value of the production in two different moments in order to explore the psychological effects caused by emergency on resident population and, consequently, on human labour.

- *Change in Gross Local Product or local valued-added:* referred to the reduction of the flows of goods and services, which can be bought and sold in markets. It can include, for example, lower output from damaged or destroyed assets and infrastructure and loss of income due to damage to marketing infrastructure.

5 Software tool

In the CRISMA project, a software tool based on the economic evaluation approach presented in section 3 will be developed to make the assessment procedure usable in practice.

Figure 3 presents the software components that will be developed as well as the interfaces within the CRISMA architecture. The software tool is to be implemented by two software components. Economic impact calculation view (in the upper part of the diagram) is a user interaction component and provides a graphical user interface (GUI) for the end users. The other component in the lower part of the diagram is a federated calculation and simulation model providing a service to calculate the costs of the crisis under investigation.

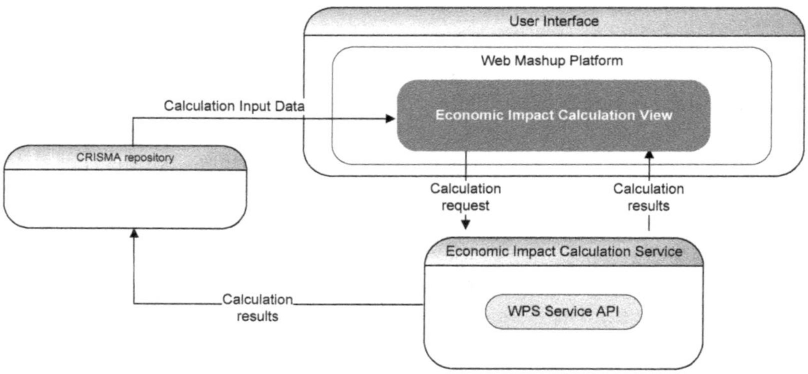

Figure 3: Software tool components.

As the diagram in fig. 3 indicates, Economic Impact Calculation View *receives some of the parameters from an external storage and some of the parameters are entered by the user. The parameters are sent to Economic Impact* Calculation Service in a calculation request which will launch the calculation and/or simulation process implemented in the Web Processing Service (WPS) instance. When the requested economic impact calculation is dispatched, the calculation and simulation service provides the results back to the user interface.

In accordance with the CRISMA architectural requirements for user interaction components, Economic Impact Calculation View is provided as HTML5 JavaScript widget. The jQuery JavaScript library and Kendo UI framework are used to build the user interface. In accordance with the CRISMA architectural requirements for federated simulation models, Economic Impact Calculation Service will be provided as Open Geospatial Consortium (OGC [20], WPS [21]). In spite of the fact that the OGC WPS standard defines how geospatial services can be executed, it can be used to describe any process and execute it by using predefined input and output parameters. PyWPS is a Python implementation of OGC's WPS 1.0.0 standard. PyWPS was chosen because it has received positive reviews and seemed appropriate for the purpose. For its simplicity and rapid deployment, Common Gateway Interface (CGI) was chosen but the Web Service Gateway Interface (WSGI) version would be recommended for its better performance to handle multiple processes [22]. The objective is that the economic impact calculation service could be implemented as a single Python process.

6 Conclusions

This paper proposes a practical approach for the economic assessment of mitigation investments for crisis management, designed to support long-term planning and decision-making. The approach covers the influence of economic impacts of mitigation measures and uncertainty on decisions. The study is based on research carried out within the CRISMA project. In the next phases of CRISMA, the approach will be further evaluated and tested. It is expected that in most cases, a different set of cost and benefit categories and decision criteria will be used without the need of any changes in the overall structure of the assessment procedure. Furthermore, a software tool based on the assessment approach will be developed to make the assessment procedure practical.

As with any empirical research, the limitations of the proposed approach must be taken into account when interpreting the results. It may not cover all the important aspects of linking vulnerabilities and damages with economic factors. Special considerations must be made regarding the expert judgement and expert panels. There are some generally recognised problems in expert judgement, such as subjectivity and overconfidence of the experts. Thus, the coverage of the expert group and competence of the experts is of prime importance.

However, the approach provides a practical structure to the integration of economic aspects of crises into the decision-making on mitigation measures. It is believed that the developed approach fulfils its intended purpose as an easy-to-apply assessment approach. Furthermore, the approach enhances the transparency of investment decision-making and contributes to the more comprehensive use of available information affecting the cost effectiveness of different mitigation investments.

References

[1] Heikkilä, A.M., Molarius, R., Rosqvist, T. & Perrels, A., Mitigating the impacts of extreme weather originated disasters by simulating the effects of different preparation and action decisions of crisis management.

Programme and abstracts of the Second Nordic International Conference on Climate Change. Adaptation research meets adaptation decision-making, eds. T. Carter, A. Perrels & M. Hildén, p. 61, 2012.

[2] Räikkönen, M., Rosqvist, T., Poussa, L. & Jähi, M., A framework for integrating economic evaluation and risk assessment to support policymakers' security-related decisions. *PSAM11 & Esrel 2012 Int. Conf. Proceedings*, Helsinki, Finland, USB memory stick, p. 18, 2012.

[3] Meyer, V., Becker, N., Markantonis, N., Schwarze, R., van den Bergh, J.C.J.M., Bouwer, L.M., Bubeck, P., Ciavola, P., Genovese, E., Green, C., Hallegatte, S., Kreibich, H., Lequeux, Q., Logar, I., Papyrakis, E., Pfurtscheller, C., Poussin, J., Przyluski, V., Thieken, A.H. & Viavattene, C., Assessing the costs of natural hazards-state of the art and knowledge gaps. *Natural Hazards and Earth System Sciences*, **13**, pp. 1351-1373, 2013.

[4] Götze, U., Northcott, D. & Schuster P., *Investment Appraisal: Methods and Models*, Springer-Verlag: Berlin – Heidelberg, 2008.

[5] Pike, R. & Neale, B., *Corporate Finance and Investment: Decisions and Strategies*, 4th ed., Prentice Hall: Harlow, 2003.

[6] Keeney, L. & Raiffa, H., *Decisions with Multiple Objectives: Preferences and Value Trade-offs*, Cambridge University Press. 1993.

[7] Boardman, A.E., Greenberg, D.H., Vining, A.R. & Weimer, D.L., *Cost–Benefit Analysis: Concepts and Practice*, 3rd ed. Prentice Hall, 2006.

[8] Fuguitt, D. & Wilcox, S.J., *Cost-Benefit Analysis for Public Decision Makers*, Quorum Books: Westport, CT, 1999.

[9] IEC 60300-3-3 ed2.0. Dependability management – Part 3: Application guide – Life cycle costing, 2004.

[10] Guha-Sapir, D. & Santos, I. (eds.), *The Economic Impacts of Natural Disasters*, Oxford UP, USA, 2013.

[11] De Groeve, T., Poljansek, K. & Ehrlich, D., *Recording Disaster Losses. Recommendations for a European Approach*, JRC Scientific and Policy Reports, 2013.

[12] McKenzie, E., Prasad, B. & Kaloumaira, A., *Guidelines for Estimating the Economic Impact of Natural Disasters in the Pacific*, The Australia Agency for International Development, pp. 1-38, 2005.

[13] Middelmann, M.H. (ed.). *Natural Hazards. Identifying Risk Analysis Requirements in Australia. Chapter 2: Impact of Natural Disasters*. Department of Industry, Tourism & Resources, pp. 7-29, 2007.

[14] Moe, T.L., Gehbauer, F., Senitz, S. & Mueller, M., Balanced scorecard for natural disasters on development in the Pacific. *Disaster Prevention and Management*, **16(5)**, pp. 785-806, 2007.

[15] Millet, I. & Wedley, W.C., Modelling risk and uncertainty with the analytic hierarchy process. *Journal of Multicriteria Decision Analysis*, **11(2)**, pp. 97-107, 2002.

[16] Tavana, M.E., Strategic alternative assessment matrix. *Journal of Multicriteria Decision Analysis*, **11(2)**, pp. 75-96, 2002.

[17] Cavasogly, H., Making sound security investment decisions. *Journal of Information Privacy & Security*, **6(1)**, pp. 53-71, 2010.
[18] Cellini, S.R. & Kee, J.E., Cost-effectiveness and cost-benefit analysis. *Handbook of Practical Program Evaluation*, 3rd ed., eds. J.S. Wholey, H.P. Hatry & K.E. Newcomer, Jossey-Bass: San Francisco, Chapter 25, 2010.
[19] Zuccaro, G., .Leone, M.F., Del Cogliano, D. & Sgroi, A., Economic impact of explosive volcanic eruptions: A simulation-based assessment model applied to Campania region volcanoes. *Journal of Volcanology and Geothermal Research*, **266**, pp. 1-15, 2013.
[20] Open Geospatial Consortium, www.opengeospatial.org
[21] Web Processing Service, www.opengeospatial.org/standards/wps
[22] Boerboom, J., Implementing the WPS Standard. A Case Study for Dissemination of Coastal and Marine Tools, Geographical Information Management and Applications, Delft University of Technology, the Netherlands, 2013.

A disaster severity assessment decision support tool for reducing the risk of failure in response operations

S. Hasani, R. El-Haddadeh & E. Aktas
Business School, Brunel University, UK

Abstract

After a disaster strikes, the disaster severity needs to be estimated in order to provide an adequate humanitarian response. The decision makers need to decide quickly about the requisite supplies and bespoke teams based on the scale and nature of the disruption. Failure of the severity assessment may impose risk to the success of the response operations, leading to the loss of lives at worst. The existing severity assessment tools employ various criteria such as intensity, frequency, vulnerability and capability. However, a framework which differentiates between the impact of disasters on communities with different coping capabilities is missing. For example, an equally intense earthquake affects Japan differently from Haiti, due to the difference in the coping capabilities of the two nations. To that end, this paper investigates the records of previous disasters to provide a holistic Disaster Severity Assessment (DSA) tool. This decision support tool accommodates physical and socio-economic impacts of the disaster on the affected population. The assessment is based on six criteria including impact time, fatality, casualty, relative financial damage, Human Development Index (HDI) and Disaster Risk Index (DRI). The resulting decision support tool may be used to diagnose the severity of the disaster immediately after it strikes. It is also capable of accommodating the imprecise data by means of a fuzzy classification system. This characteristic allows the decision maker to draw a realistic picture of the disaster response required based on the affected population capabilities and reduce the risk of failure in disaster response operations.
Keywords: disaster response operations, severity assessment, disaster risk management, risk reduction, decision support, fuzzy classification.

1 Introduction

Disaster as a physical disruption leaves a community incapable of coping with the impacts of the disaster. Therefore the timely supply of adequate aid and supplies during the response operation would save lives. Conversely, the risk of failure of the disaster response operation may be increased by inaccurate assessment of the disaster severity.

The present paper suggests that a decision support tool drawn upon the existing data from previous disasters could be used as a guideline for the severity assessment of future disasters with similar characteristics. This scenario-based decision support technique would assist in estimating the impact of the disaster on the affected community in the early stages of the disaster and provide an accurate estimation of the required aid at an early stage.

The impact of disasters has been the subject of a number of studies providing a scalable framework for classifying any given disaster's disruption effect. These frameworks are based on, amongst others: intensity, affected population, temporal and geographical dispersion [1–3]. Other approaches have been developed to take into account the social and cultural variables [4–6] as well as the socio-economic variables such as coping capability, vulnerability, social preparedness, human development indicator, disaster risk index, exposure, resourcefulness and demand/supply capability of the affected communities [7–13]. The significance of these approaches is to differentiate between the actual impacts of the disaster on various communities. For example, an intense flood, which barely puts lives and infrastructure in danger due to its occurrence in a remote area, might not require a humanitarian response [8]. Also two disasters with a similar disruption magnitude could have a low impact in a community with more coping capabilities such as Japan, whilst it might create a catastrophic impact on Haiti with a lower level of coping capabilities.

One problem that decision makers are facing is that, despite the existence of the appropriate measures in the literature, a holistic severity assessment tool, which differentiates between the various scenarios of disasters, is yet to be developed. Another problem is that, due to the time pressures during a disaster, most of the data remain un-obtainable, unreliable or imprecise. Therefore the decision tool needs to be based on existing or obtainable data and have the capability to accommodate imprecise values. To that end, the current study aims to design a disaster severity assessment tool based on previous experience; this tool would be quick, easy to use and capable of working with existing data. It is also required to accommodate the imprecise values inherent in an uncertain disaster situation.

2 Disaster severity assessment method

The first step is to develop a method for the classification of previous major disasters. The aim is to identify a pattern for disaster severity. In this step, the secondary data from four various sources are obtained. Primarily, the prominent natural disasters occurring after 1980 mentioned in the encyclopedia of disasters by Gunn [14] have been analysed. The results were compared to the costliest/deadliest disasters in NatCatSERVICE [15] provided by Munich Reinsurance Company and containing 26,000 natural disasters since 1979. These data were then compared to the EM-DAT database [16] published by the Centre for Research on the Epidemiology of Disasters and are based on 17,000 natural and technological disasters since 1900. Finally the results were compared to the DesInventar project [17] and resulted in identification of 61 disasters as the data source for the present paper (table 1).

Table 1: An example of the data source for the present paper.

Year	Disaster type	Origin	Fatalities	Affected people	Overall Loss
1983	Earthquake	USA	1545	31	Not specified
1985	Earthquake	Mexico	9500	2130204	4104
1988	Earthquake	Armenia	Non-existent in the data base		
1989	Earthquake	USA	62	3757	5600
1990	Earthquake	Iran	26796	267628	500
1994	Earthquake	USA	60	27000	30000
1995	Earthquake	Japan	5297	541636	100000
1999	Earthquake	Taiwan	2264	108664	14100
1999	Earthquake	Turkey	17127	1358953	20000
2001	Earthquake	India	20005	6321812	2623
2001	Earthquake	Peru	145	349978	300
2003	Earthquake	Iran	40000	710000	8000
2005	Earthquake	Iran	612	94766	80
2005	Earthquake	Pakistan	73338	5128309	5200
2008	Earthquake	China	87476	45976596	85000
2011	Earthquake	New Zealand	181	301500	15000
2012	Earthquake	Italy	17	14350	Not specified
2004	Earthquake	Japan	89	84792	2300
2004	Tsunami	Sumatra	80	5063	500
2010	Tsunami	Chile	562	2671556	30000
2011	Tsunami	Japan	19846	368820	210000
2004	Tsunami	Indonesia	165708	532898	4451.6
1991	Flood	China	1729	210232227	7500
1993	Flood	China	1000		6061
1993	Flood	USA	48	31000	12000
1996	Flood	China	2775	154634000	12600
1998	Flood	China	3656	238973000	30000
1998	Flood	North Korea	50	10172	Not specified
2002	Flood	Europe	167	49500	500
2008	Flood	USA	24	11000148	10000
2010	Flood	Pakistan	1985	20359496	9500
2011	Flood	Thailand	813	9500000	40000
2006	Mudslides	Philippines	1399	2562517	66.4
2004	Cyclone	USA	10	30000	16000

The second step is to develop the disaster severity assessment (DSA) tool. The present study combines the De Boer/Ferro classification models and customises the combination to suit its purpose. De Boer [18] and Ferro [3] provide a model for classification of disasters from the medical perspective. Their model ranks the disasters based on scores associated to each value including cause of disaster, national or international assistance, temporal, geographical and physical affected area, the number of fatalities and casualties, rescue time and severity of injuries (table 2).

This model has been further developed to accommodate the socio-economic factors of the affected community. The resulting DSA framework is exhibited in table 3. The framework shares three characteristics with DeBoer/Ferro framework (time, fatality and casualty) and adds three new terms, i.e. financial damage, human development index (HDI) and disaster risk indicator (DRI). These terms are explained in the following section.

Table 2: The DeBoer/Ferro severity framework.

Classification	**Grade**	**Score**	**DeBoer**	**Ferro**
Cause	Natural/Man-made	1/0	+	
Effect on surrounding community (National assistance/ international assistance)	Simple/Compound	1/2	+	
Impact time	> 24 hour/1–24 hour/1 hour >	2/1/0	+	+
Radius of disaster area	10 km </1–10 km/< 1 km	2/1/0	+	+
Number Causalities	25–100 causalities alive or dead or 10–50 admission to hospital/ 100–500 causality or 50–250 admission/more than 500 causalities of more than 250 hospital attendance	0/1/2	+	
Rescue time (rescue first aid + transport + evacuation)	24 hours </6–24 hours </6 hours	2/1/0	+	+
Effect of infrastructure (impact site + filter area)	Simple/compound	1/2		+
Number of dead	< 100/100 <	0/1		+
Number of injured	1000 < 100–999 < 100	2 < 1 < 0		+
Average severity of injuries sustained	2 < 1–2 < 1	2 < 1 < 0		+
Total	**1–13**			

Each category is then defined by a number of thresholds starting from the minimum value observed in the disaster records (for example 1 day impact time) for the lower bound and ending to the upper bound where the outliners are observed (for example more than 2 weeks). The categories for impact time, fatalities, casualties, financial damage and severity are designed in a way that each category exhibits at least ten frequencies in the records. For the last two terms (HDI and DRI), the categories are identical with the thresholds used in their original reports [12, 13]. The first characteristic is impact time associated with the number of days in which the disaster is in effect. The second characteristic is the damage to the population by either loss of lives (fatalities) or other types of damage including injuries (casualties). The third characteristic is the 'relative financial damage' defined in this study as an indication of the material losses of the affected community. This characteristic is originally expressed in million dollars in various sources (NetCatService, EM-DAT). However, 1 million dollar worth of damage in a low-income country, such as

Table 3: Disaster severity assessment framework.

Characteristics	Categories	Linguistic value	Numeric value
Impact time (day)	Over 2 weeks/ between 1–2 weeks/ 1 day–1 week/1 day	Extreme/long/ medium/short	3/2/1/0
Fatalities (Person)	Over 1,000/ between 100–999/under 99	High/medium/ low	2/1/0
Casualties (Person)	Over 1,000,000/ between 1,000,000–99,999/ below 99,999	High/medium/ low	2/1/0
Relative financial damage (USD)	Over 10,000/between 1,000–10,000/ between 1,000–100/under 100	Extreme/high/ medium/low	3/2/1/0
HDI (%)	Under 4.66/ between 4.67 and 6.40/between 6.41 and 7.59/ over 7.59	Low/medium/ high/extreme	3/2/1/0
DRI (%)	Under 3.65/between 3.66 and 5.72/between 5.73–7.44/ between 7.45 to 10.58/over 10.59	Low/medium/ high/very high/ extreme	0/1/2/3/4
Total Severity	Less than 5/between 5 and 10/more than 10	Severe/very severe/extreme	

India where the GDP/capita in 2005 was $740 is likely to cause deeper socio-economical damage than in a high-income country such as USA with $42,516 GDP/capita in the same year. To that end, the relative financial damage used in the DSA is calculated using the following formula:

$$\text{Relative financial damage} = \frac{\text{Original damage of disaster}}{\text{Affected country GDP per Capita}} \tag{1}$$

The result expresses the relative financial damage of a specific disaster in a specific year.

The last two categories indicate the capacity of the affected country for coping with disaster including the HDI and DRI. The human development index [12], published by UNDP, is based on health, education and living standards for each country. Where HDI is the lowest (such as Haiti), the country is less likely to have the capability to cope with the damages comparing to a county with higher HDI (such as Japan).

Another characteristic is the disaster risk indicator, which calculates where the high exposure to natural hazards and climate change coincides with very vulnerable societies. This index is published in the World Risk Report [13]. Therefore where DRI percentage is higher, it is more likely that the affected society will fail to cope with the situation.

The present Disaster Severity Assessment classifies disasters into 3 categories: severe disasters (less than 5 DSA rank), very severe disasters (rank between 5 and 10) and extremely severe disasters (rank above 10). Due to publishing limitations, just one example of the results obtained from the DSA technique is exhibited in table 4.

Table 4: Example of DSA result.

Severity	Disaster	Impact time (Day)	HDI (%)	DRI	GDP/ capita	Relative financial damage	DSA rank
Severe	2012 Italy earthquake	1	V	112	33049	484.13	**3**
	1982 Mexico eruption	1	H	94	2366	49.45	**5**
Very Severe	1993 USA flood	60	V	127	25327	473.8	**6**
	2004 Haiti cyclone	2	L	21	401	124.69	**9**
Extreme	2008 China earthquake	1	M	78	3414	24897.48	**12**
	1996 China flood	27	M	78	703	17923.19	**15**

The significance of this framework is that it could be used as a quick guideline for decision makers under time pressures at the early stages of a disaster. By knowing the approximate number of fatalities, casualties and the financial damage, the disaster severity can be estimated at any given time and the response actions could be conducted accordingly. Also as the disaster develops, the numbers and the model can be easily re-adjusted. For a more technical decision maker that prefers to work with mathematical decision-making models, the DSA data could be used as the basis for building the fuzzy system's knowledge base to diagnose the severity of the situation in linguistic or numerical terms. An example of such system design is the subject of the next section.

3 Fuzzy classification system

According to Cintra *et al.* [19], classification is an important task in decision-making where it provides a key to the interpretation of the system's behaviour. But in some cases the parameters and input data are imprecise and the framework is required to process adequately the imprecise data as well as the associated uncertainty. To that end, the fuzzy technique introduced by Zadeh [20] is selected for this study because of its capacity to classify the linguistic attributes associated to disaster situation. This classification is based on the rules drawn from a particular knowledge base, which, in this paper, are the result of the presented DSA in the previous section. The simplified steps of the fuzzy classification employed in the paper are described below:

1. Consider the pattern that requires to be classified (here it is the severity pattern of the disaster). The meaning of the linguistic terms is defined by their membership functions

$$\mu_S: S \rightarrow [0,1] \tag{2}$$

where *S* stands for Severity. For example, the membership function of the output (here Severity) is expressed in three linguistic terms including severe, very severe and extreme. A similar set of membership functions are required to be generated for each input (here including impact time, fatality, DRI, etc.).

Therefore, a simple presentation of this concept for severity can be seen in fig. 1.

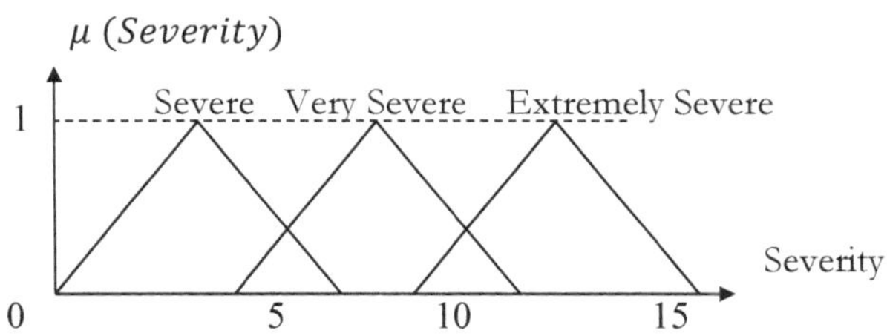

Figure 1: The graphic representation of the Severity membership function.

2. A set of fuzzy rules based on previous experience or expert opinion is defined to exhibit the relationships between inputs or terms (here the impact time, casualty,) and output (here severity). The rule base can be defined as $\{Ri \mid i = 1,2,n\}$, therefore the (semi) qualitative rule can be defined as (3):

$$if\ x_{i1} is\ A_{i1}\ and\ x_{i2}\ is\ A_{i2}\ , \ldots, and\ x_{in}\ is\ A_{in,}\ Then\ y_{i1}\ is\ B_{i1} \tag{3}$$

where *x* is the input (here impact time, fatality, direct), *y* is the output (here severity) and *A* and *B* are the linguistic variables such as high, low, extreme. Since in this model the impact time can take four values (Extreme, Long, Medium, Short), fatality and casualties could each take three values (High, Medium, Low), Relative financial damage and HDI could take four values (Extreme, High, Medium, Low), DRI could take five values (Extreme, Very High, High, Medium, Low) and finally severity could take 3 values (Severe, Very Severe, Extremely Severe), the number of possible rules are 4×3×3×4×5×3 = 2160 rules. However the review of the DSA result in the previously recorded disasters (61 disasters) shows that the rules followed in reality can be summarized into the sum of 48 rules including the following examples.

Box 1: Rules derived from historical data.

Rule (1): if fatality is High and casualty is High and impact time is Extreme and HDI is Medium and DRI is Medium and financial damage is Very high then severity is Extreme.

Rule (48): if fatality is Low and casualty is Low and impact time is Medium and HDI is Very high and DRI is Very high and financial damage is Low then severity is Severe.

3. The fuzzy inference system runs the pattern through the rules to identify the class in which the output (severity) belongs. The compatibility of the rules will then be compared to the pattern of severity in disaster. The class predicted by the rule is assigned to the attributes. Fig. 2 exhibits how the severity of disaster can be diagnosed by identifying the conjunction between the imprecise variables' space. For enhancing the graphical illustration, a two-dimensional pattern space [0, 1]×[0, 1] may be assumed by having the variables of fatality and DRI. Based on the following simplified rules, fig. 2 can be illustrated as follows.

Box 2: Simplified rules for a two dimensional space.

If fatality is high and DRI is medium, then severity is extreme And If fatality is low and DRI is low then severity is severe

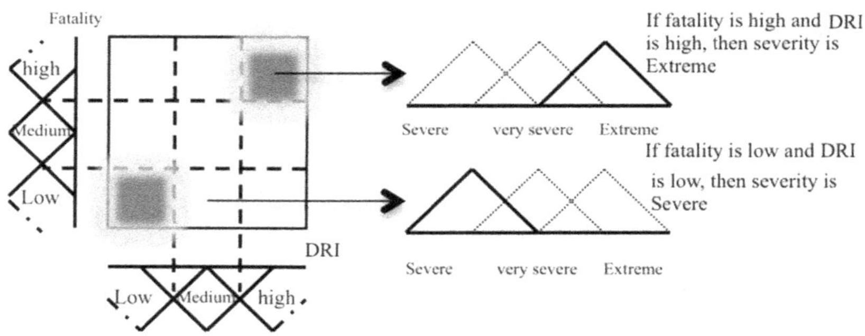

Figure 2: Overlapping areas of fuzzy sets.

In a more complex space with more dimensions, if there are K fuzzy subsets $\{A_1^k, \ldots, A_k^k\}$, a triangular membership function [21] can be calculated based on the formula

$$\mu_i^k\,(x) = \max\{1 - |x - a_i^k|b^k, 0\} \quad , i = 1,2, \ldots, k\ (k \geq 2) \qquad (4)$$

where (a) and (b) are, respectively, the upper and lower bound in each categories of input (table 1).

Formula (4) can be translated to the following *if-then* construct in MS-Excel to calculate $\mu(x)$ of the x-values (impact time, DRI, etc.) and $\mu(y)$ of the y-value (severity):

IF (AND (x > *left*, x < *right*), 1 – ABS ((*left* + *right* – 2x)/(*right* – *left*)), 0) (5)

where *left* and *right* represent, respectively, the lower and upper bounds of the threshold (table 1) and ABS is the absolute value. Part of the result is exhibited in table 5.

Table 5: An example of the data used for triangular membership functions.

Term		Position	*x*-value	μ(*x*)	Term		Position	*x*-value	μ(*x*)
Fatalities	High	*Left*	1000.00	0.00	HDI	Low	*Left*	0.00	0 00
		Right	10000000.00	0.00			*Right*	4.66	0.00
	Medium	*Left*	100.00	0.00		Medium	*Left*	4.67	0.00
		Right	999.00	0.00			*Right*	6.40	0.93
	Low	*Left*	0.00	0.00		High	*Left*	6.41	0.00
		Right	99.00	0.00			*Right*	7.58	0.28
Severity	Severe	*Left*	0.00	0.00		Extreme	*Left*	7.59	0.27
		Right	5.00	0.00			*Right*	100.00	0.00
	Very Severe	*Left*	5.00	0.00					
		Right	10.00	0.00					
	Extreme	*Left*	10.00	0.00					
		Right	20.00	0.00					

The data provided in table 5 and the two simplified rules employed in fig. 2 were used for simulation and control of the model with an open source computer program (Fuzzylite, 2013) as is exhibited in fig. 3.

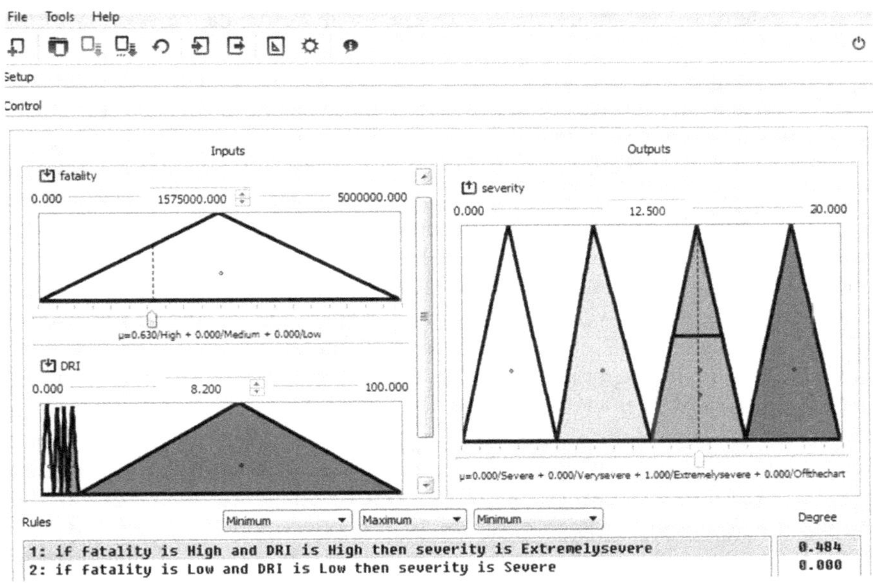

Figure 3: A snapshot of the created fuzzy model based on the data in table 3.

Fig. 3 employs the clipping as one of the most common ways of showing correlation between the rule outputs with the truth-value of the rule input [22]. It basically cuts the output membership function at the level of the input truth. Slicing the top of the membership function leads to the loss of part of the information in a fuzzy set. In the above example, the graphs show that where the fatality is equal to 1,575,000 persons (high) and DRI is 8.2% (high), the severity is 12.5 (extremely severe) when rule 1 is activated with 0.484 degrees of strength. This principle can apply to any possible values of the outputs and inputs.

4 Conclusions

This paper is an attempt to classify the possible scenarios of disaster impact in order to diagnose the severity of the disaster in its early stages of development when information is most scarce. In this situation, the decisions need to be made under time pressures and with the minimum amount of primary data. Therefore a decision support tool is required to quickly diagnose the severity of the situation based on the data available. To that end, in this paper, a decision support tool is designed targeting two groups of decision makers. The first tool (DSA) is a simple framework based on previous disasters and the socio-economic characteristics of the affected community. It is constructed from a simple chart that can be used by any decision maker without mathematical background. Another significance of this framework is that its linguistic values and boundaries could be altered according to the situation. It enables the decision makers to narrow the various classes of severity and estimate the severity of the situation with more precision if required.

The paper also explains the preliminary steps to simulate the DSA in a fuzzy classification system which can be used by a more technical decision maker and accommodate imprecise values and map the disaster severity. The result could assist the decision makers to address more accurately and diagnose the severity of the situation and reduce the risk of failure in disaster response operation. The study also addresses the issue of data limitations. First, the data from various sources show discrepancies. In many cases the data provided in the Munich RE and EM-DAT databases and the NetCatService database report different statistics. Although this weakness is partially addressed by the use of fuzzy logic, which is designed for imprecise data, it is noteworthy to remember the inconsistency of the reported data. Also the processed data is drawn from 61 natural disasters with the most destructive magnitudes. Although these amounts of data are sufficient for statistical interpretation, more data need to be processed for building a rich knowledge base for defining fuzzy rules. The data also exclude man-made disasters. However the results to some extent are extrapolatable to any types of disasters. This study is part of an extensive on-going research, which attempts to reduce the risk of failure in humanitarian response networks by developing a scenario-based decision framework for disaster partner configuration.

References

[1] Rutherford, W.H. & De Boer, J., The definition and classification of disasters. *Injury: British Journal of Accident Surgery*, **15(1)**, pp. 10-12, 1982.

[2] Gad-el-Hak, M., *Large-Scale Disasters: Prediction, Control and Mitigation*, Cambridge University Press: London, 2007.

[3] Ferro, G., Assessment of major and minor events that occurred in Italy during the last century using a disaster severity scale score. *Pre-Hospital and Disaster Medicine*, **20(5)**, pp. 316-323, 2005.

[4] Carr, L.J., Disaster and the sequence-pattern concept of social change. *American Journal of Sociology*, **38(2)**, pp. 207-218, 1932.

[5] Barton, A.H., *Communities in disaster: A Sociological Analysis of Collective Stress Situations*, Doubleday: Garden City, NY, 1969.

[6] Britton, N.R., Developing an understanding of disaster. *Australia & New Zealand Journal of Sociology*, **22(2)**, pp. 265-271, 1986.

[7] Rongier, C., Gourc, D., Lauras, M. & Galasso, F., Towards a performance measurement system to control disaster response. *Collaborative Networks for a Sustainable World, IFIP Advances in Information and Communication Technology*, **336**, pp. 189-196, 2010.

[8] Wickramaratne, S., Ruwanpura, J., Ranasinghe, U., Walawe-Durage, S., Adikariwattage, V. & Wirasinghe, S., Ranking of natural disasters in Sri Lanka for mitigation planning. *International Journal of Disaster Resilience in the Built Environment*, **3(2)**, pp. 115-132, 2012.

[9] Mishra, V., Fuloria, S. & Bisht, S.S., Enhancing disaster management by mapping disaster proneness and preparedness. *Disasters,* **36(3)**, pp. 382-397, 2012.

[10] Berren, M., Beigel, A. & Barker, G., A typology for the classification of disasters: Implications for intervention. *Community Mental Health Journal*, **18(2)**, pp. 120-134, 1982.

[11] Eßig, M. & Tandler, S., Disaster response supply chain management (SCM): Integration of humanitarian and defence logistics by means of SCM. *Economic Impacts of Crisis Response Operations – An Underestimated Factor in External Engagement*, eds. W. Feichtinger, M. Gauster & F. Tanner, Austria: Institut für Friedenssicherung und Konfliktmanagement (IFK), pp. 283-310, 2010.

[12] Human Development Reports (HDR), United Nations Development Programme (UNDP) online, http://hdr.undp.org/en/statistics/

[13] World Risk Report 2012, online http://www.worldriskreport.com/uploads/media/WRR_2012_en_online.pdf

[14] Gunn, A.M., *Encyclopaedia of Disasters: Environmental Catastrophes and Human Tragedies*, Greenwood: Connecticut, 2007.

[15] Significance natural disasters since 1980, Munich, reinsurance company, Online, http://www.munichre.com/en/reinsurance/business/non-life/georisks/natcatservice/significant_natural_catastrophes.aspx

[16] The international disaster database (EM-Dat), The Centre for Research on the Epidemiology of Disasters (CRED), online http://www.emdat.be
[17] Inventory system of the effects of disasters (DesInventar), Online http://www.desinventar.org/en/database
[18] De Boer, J., Definition and classification of disasters: introduction of a disaster severity scale. *The Journal of Emergency Medicine*, **8**, pp. 591-595, 1990.
[19] Cintra, E., Estevam, H., Hruschka, R. & Nicoletti, M., Automatic construction of fuzzy rule bases: a further investigation into two alternative inductive approaches. *Journal of Universal Computer Science,* **14(15)**, pp. 2456-2470, 2008.
[20] Zadeh, L.A., Fuzzy sets. *Journal of Information and Control*, **8**, pp. 338-353, 1965.
[21] Ishibuchi, H., Nozaki, K., Yamamoto, N. & Tanaka, H., Construction of fuzzy classification systems with rectangular fuzzy rules using genetic algorithms. *Fuzzy Sets and Systems*, **65**, pp. 237-253, 1994.
[22] Negnevitsky, M., *Artificial Intelligence*, Addison-Wesley: Harlow, England, 2005.

Just and safe housing for informal settlers in the developing world

J. Cuadra, M. Samples, R. Brower & J. Dilling
Florida State University, USA

Abstract

The risk of disaster is increasing globally, particularly in under-developed or developing countries, due to population pressure and poverty. This risk is elevated where informal and non-permanent settlement housing is prevalent as these communities are caught up in a vicious repeat cycle of destruction and loss of life due to their location in high risk areas. In this study, the focus is principally on the actions of informal settlers who inhabit areas that leave them vulnerable to disasters and, the often exasperating efforts of government leaders attempting to relocate them to safer settings. The authors believe that governments (the state) and communities often have difficulty finding a point of agreement because they have varying rationales on what is "just" and what constitutes safe and appropriate housing. These perspectives are founded on different forms of judgment: one based on scientific facts and politics and one on social, emotive and aesthetic experiences. Practitioners often fail at solving these difficult problems because, during dialogues, people tend to only communicate positions rather than interests. A better understanding of the feelings and emotions of those affected could lead to more accurate transmission of their own positions and interests to others. In this study, an attempt is made first to clarify the origins of the conflicts between both perspectives using relevant literature on the topic; to later make use of preliminary data from the authors' field work in the Philippines to serve as an example of the benefits of a careful study of the cultural, social, economic and political factors and interactive dynamics that helps explain the way people decide where to live. The collection and analysis of such data for different case studies should contribute towards a better understanding of the individuals at risk and formulation of more effective disaster reduction strategies.
Keywords: high risk settlements, justice, relocation, community resistance.

1 Introduction

More than a billion people live in marginal and informal settlements, many without access to basic services and, very frequently, in high-risk areas. The risk of disaster is increasing globally, particularly in under developed or developing countries, due to population pressure and poverty. This risk is elevated where informal and non-permanent settlement housing is prevalent. Their vulnerability to disaster events is often compounded by a lack of infrastructure, environmental degradation and increasing challenges exacerbated by climate change and sea level rise. Researchers and local governments are in need of a better understanding of the societal and community dynamics of these groups in order to provide for permanent housing and a place they can call home. The understanding of human settlement vulnerability, when taken together with the underlying factors that create and maintain settlements, is crucial in making settlements more resilient.

2 Conflict between justice and safety

In this study, the focus is principally on the actions of informal settlers who inhabit areas that leave them vulnerable to disasters and, on the other hand, the often exasperating efforts of government leaders attempting to relocate them to safer settings. As scholars dedicated to implementing formal systems to prevent and respond to human calamities the authors largely share a mindset with the government leaders. On the other hand, they are mindful that people choose to live where they live for many reasons; many of these are deeply embedded. The state and its officials are then faced with the necessity to balance between ensuring safety and providing the informal settlers just treatment. But why should there be worry about justice in the context of high-risk settlements? Fainstein [2], in her work on justice in the urban planning context, suggests that, not only governments, but specifically the planning profession, have neglected the needs of the powerless. Currently the critiques of planning and public policy focus on the wrong actions of the centralized decision making process, those that deliberately block the voices of the affected publics. Urban development has a history of inequality and policy implementation that exacerbates the disadvantages suffered by low-income minorities. This is especially true for those living in informal settlements.

Relocation or involuntary resettlement is often deemed the easiest solution to the problem of vulnerable communities. From the state point of view, there are good reasons to do this: to protect vulnerable groups and to reduce fiscal drain resulting from multiple relief and reconstruction efforts. However, in the developing world, relocation by the government is often a failure. The principal reason for this failure is "underweighting the welfare of the population as a criterion for the selection of the relocation site" [3]. Welfare in this sense is understood as the compilation of factors that create wellness and sustain human settlements beyond economy and geographic location. It is important, therefore, that a concerted effort is made to understand the life-world of those on the other side of the equation. It can be imagined then that for the informal settler justice means being able to choose where to live. This includes having access to goods and services even if it comes with the probability of suffering the effects of natural phenomena. On the other hand the state perceives justice in this context as the equal opportunity to enjoy safe and stable housing. Fainstein [2] tells us that a general understanding of justice in urban context should be separated from the concept of "good life" since the definition of "good" varies from person to person. Then, she proposes that justice in the urban context consists of actions and outcomes that comply with elements of equity, democracy and diversity. By complying with these three elements of justice a fairer process and outcomes can be ensured for the least advantaged. Informal settlers threatened with relocation perceive state treatment as unjust, whereas state actors believe they are contributing to some extent to a just cause whether the public understands it or not. Scott's [4] perspective reflects how morality and perceptions of justice play into the decisions of informal settlers, who often feel they have a moral right to occupy these properties, since government and societal elites have previously stripped them of traditional communal or usufruct property rights elsewhere.

But communities living in informal settlements are not always defenseless and passive. Organized groups have the potential to influence state legislation. According to Friedmann [5], society is the handy work of the people and the state cannot dominate organized people, at least not for too long. Interestingly communities facing relocation find strength in constant dialogue and collective critical analysis of their situation. Habermassian theories suggest that by use of constant dialogue communities are able to communicate in public arenas, sort out their priorities and decide courses of action [6]. These public arenas or areas of social interactions such as bars, churches, cafes or markets are an important asset to peasant resistance according to Scott [7] since they represent focal points of autonomous public life.

Faced with imminent relocation, communities are pushed to develop strategies to fight back against state intentions. Scott [8] argues that these acts of resistance and thoughts about resistance are in constant communication; it is a cycle of trial and error by which communities develop over time effective strategies to exert pressure. The case of Latin America is a good example of the slow but steady self-empowerment movement among informal settlers. In the past, informal areas were ignored and bulldozed, but citizen pressures on politicians and public protests have secured them a means for upgrading infrastructure and legalization of their communities [9].

Ideally, governments and the politically powerful should ease the process of adaptation for those being resettled. Also, "it must make sure land is available that is not at risk". Land prices dependable on income lead to communities settling in ill-suited areas [9]. In reality, political interests can harm communities; e.g., in developing countries in Latin America, governmental support is provided to neighborhoods sitting in risk areas in exchange for electoral votes. Both electoral seasons and natural disasters represent an important window of opportunity not only for policy makers to enact legislation and improve zoning, but also for settlers to demand better conditions and opportunities from the politically powerful.

Fainstein [2] argues that a possible solution for the conflicting concepts of justice and safety is to include urban planners as mediators in the conflict between state and informal settlers so as to bring to public attention certain topics that may affect citizens. This planner would distribute pieces of information crucial to the cause for resistance. This is an advocacy planner. However, this could also be the role of emergency managers. Unfortunately, state officials, urban planners and emergency managers must also comply with norms of ethics and loyalty to the ultimate employer which in most cases is the state. Communities living in informal settlements have strong sentiments on and interest in these causes, having these government officials on their side would be crucial to real social movement. Planners and emergency managers are confronted, then, with the certainty that contribution to these social causes could produce significant outcomes closer to the ideal of justice.

Non-governmental organizations (NGOs) on the other hand also have the potential to work closely with informal settlers and build participative plans to accommodate their social and economic concerns. Many of these NGOs also possess the advocacy and interpersonal communication skills to negotiate with government leaders and ameliorate their apprehensions as well. These NGO actors are included in the roles of "brokers" and "translators" [10]. But their activities also reflect significant structural features of their relationships with government leaders and informal communities. That is, NGO actors' positions within networks affect their abilities to shape others' thoughts and actions as some actors are largely under the influence of others and constrained from taking novel actions [11]. In addition, their abilities to shape others' thoughts and negotiate on behalf of informal settlers hinge in part on cultural capital [12] and class similarities they share with government officials.

In acknowledging the underlying dynamic of relocation resistance, there is no wish to romanticize these activities. The authors believe, however, that the underlying sentiment of informal settlers' "moral economy" must be accommodated in any planning that government leaders undertake in working with these settlers. The settlers' sense of moral entitlement ought to be sought in gathering data from them and government's eventual disaster planning must anticipate the practical implications of informal settlers' sentiments.

3 Opposing rationalities: state and informal settlers

What is dividing state and informal settlers' perspectives on safety and appropriate living conditions? Friedmann [5] argues that in common language an action is deemed rational when it conforms to certain forms of socially accepted judgment, such as economic efficiency or conforming to socially expected behavior. Socially accepted judgment and behavior are susceptible to changes among different countries, cultures, social levels and time periods. Rationality could then potentially be defined differently by certain people and affected by context, culture and social norms.

Let the focus be on the context of informal settlers and their decisions to reside in areas of demonstrated risk. State officials usually base their rationale on scientific analysis: equating levels of risk by combining vulnerability and hazard exposure. The result of this simple equation provides government officials with a scientifically accepted judgment to propose relocation. Unfortunately, this is a strictly scientifically accepted judgment not always compliant with socially accepted judgment. Conversely, the community's reason for living in such areas belongs to a different category of judgment, one characterized by considerable emotional and social content.

It could be argued that the rationale applied by the state is narrowly constructed around scientific analysis and the specific political pressures on government actors to take decisive action. Similarly, informal communities have constructed narrow realities around their own everyday social experiences. The state rationale may fail to comprehend complex situations as it tends to simplify real world scenarios. Forester [13] tells us about some of the failed assumptions from the state:

- overall comprehension of the problem,
- possession of the definite list of possible alternatives to consider,
- absolute understanding of the potential consequences of each alternative,
- full understanding of the values and preferences of citizens,
- adequate time, skills and resources to solve the problem.

The community rationale is not exempt from distortions either. Socially, emotionally or aesthetically generated knowledge is subject to constant change and is also difficult to measure for comparative purposes. Attaching a numerical value to these qualities is nearly an impossible task. The rationales of members of the informal community may appear similarly illogical from the vantage point of state actors, and these rationales distill easily to a single viewpoint that represents the inhabitants uniformly. Their variability reflects the diverse physical locations of their dwellings, the varieties of informal employment in which they are engaged, their particular family and other social relationships, and the varied daily routines of their family members. Thus, among the residents of the same informal settlement, there may be disagreement about the social and emotional value of various community assets and how much they would be willing to sacrifice to protect these assets. Here, the reason why state and communities do not seem to reach agreement is observed: their rationales are founded on different bases of judgment. It is then as if they were speaking in two different languages, and these languages reflect the particular features of their distinctive social embeddedness [14]. Often practitioners fail to solve difficult problems because during dialogues people tend to only communicate positions rather than interests [1]. There is value in finding a common ground of understanding. That might be through expression of interests rather than figures and facts, on the one hand, or socially-embedded experiences and emotional attachments, on the other. It is important then to differentiate between the judgment foundations for each perspective.

4 Exploring informal settlers' rationality

Why do people choose to live in these areas of demonstrated risk and vulnerability? They may make that choice because of land tenure; social and cultural ties to the community; availability of fertile land and food sources; and access to transportation and commerce. For the urban poor, it is often due to the availability of affordable housing and proximity to livelihood opportunities [3]. A look back at the history of human civilization is instructive for the reasons people decide to settle in certain areas over others. These early communities were just starting to discover agriculture, wild fruit harvesting, animal domestication and fishing as well as shelter construction techniques. These important advances changed and influenced their relationship with the land they inhabited. By attaching themselves to this new land that provided them with food and shelter the first human settlements embraced a sedentary society. And so, the ideal pattern of settlements arose to include access to food and other materials in commune with others with similar aspirations. Still, today every person needs access to the products of the land with the benefits and services of towns and cities. "The ideal pattern is one which equalizes the opportunities of town life and country life to the maximum extent" [15].

It is rational to assume that humans will continue to gather around areas with more resources. The consequence of such heavy clusters around resources is increased vulnerability to natural and man-made hazards – the "on-going societal and man-environment relations that prefigure [disaster]" [16]. In the past, the amount of damage in cities was reduced because of two factors: 1) the most dangerous sites were avoided and 2) cities were relatively small and less densely populated. Today our cities are larger, leading some communities to occupy high risk areas [17]. It follows that those with lower income tend to populate the higher risk areas because of the reduced value of the land and the minimum legal requirements to settle. Informal communities often settle on public land deemed unsuitable for development because of its inherent risk. They often cannot afford to live elsewhere. These sites also lack risk-reducing infrastructure and services, therefore, the inhabitants get less support for reconstruction after a disaster. Large numbers abandon poor rural regions because of the lack of opportunities and the appeal of distant cities. This usually means moving from a rural home to a city slum. Comparatively, at the global level, the gap between the rich and the poor has widened following the path of technological development. It could be assumed that the pattern of settlement is directly affected by the advances in technology and the disappearing trend of middle class dwellers. These forces are pushing people into the big towns and cities directly into the line of danger.

4.1 Cultural and emotional attachment to at-risk areas

The place that we call home is not just a point in a geographical setting. Communities develop shared values and shared identity with place [18]. Wellbeing and a sense of comfort can be associated with familiarity with place; disasters destroy that sense of security. Post-disaster survival stories often include reports of unrecognizable surroundings and loneliness. To lose connection with the place where we live means to feel abandoned by the people that form part of our daily lives.

Communities that possess a rich heritage consider it a serious matter to conserve their traditional ways of life even after a disaster. Resettled groups report difficulties in maintaining a traditional way of life and traditions that can be successfully transmitted to the new generations. The Moken people from Thailand argue that after the tsunami that destroyed their houses, their children lost interest in their own language and culture. The elders blame this on the location of the rebuilt community near a road that links them to city areas [19].

The town of Yungay in Peru is a good example of survivors willing to return to a devastated area. The great earthquake of 1979 followed by an avalanche from Mount Huascaran buried the town. Survivors claimed that they were sons of that devastated land and therefore were accustomed to dying and losing family. These strong remarks caught the attention of researchers like Oliver-Smith [20] and Bode [21]. Oliver-Smith [20] concludes that "The determination of the survivors to found and maintain a new Yungay, albeit one of vastly different appearance, constitutes a refusal to allow part of their identity and their culture to die". Similarly, Bode [21] reported that "it was hoped that the streets, too, could remain in their places and could retain their own names"; survivors also expressed "we would be very sad and confused if streets were called by other names". In his report for the UNISDR, Satterthwaite [9] argues that communities at risk that refuse to relocate have an option to stand up to government policy by developing proposals to show how to survive in such areas. There are cases where the location of a community shapes the way people associate and solve their problems; one can even argue that such conditions stimulate citizen participation and creativity.

4.2 Economy influencing informal settlers

Being part of a community not only benefits an individual's inner well-being, but also has powerful effects on the economic aspects of their life. Groups with limited resources tend to cluster in groups of self-support surviving by surrounding themselves with helping neighbors and family members. Similarly, disadvantaged groups such as women and the elderly find support in community unity; LaFerrara [22] argues that "people who do not have access to the formal labor market and whose options in the informal market are unattractive can often benefit from pooling resources and working in groups". Several examples in Latin America illustrate that uneducated women and single mothers benefit greatly from associating in small cooperatives which can get funding to start micro businesses in either the manufacture or service areas. Thus, it can be inferred that there is a higher chance of alleviating poverty through community associations and housing stability. For these groups, the community or neighborhood is more than a geographical point as it represents a place with higher opportunities for survival. Often, a lack of affordable land near impacted settlements occasion relocation to areas distant to the original settlement where land may be available and less expensive. Relocation on the grounds of potential or eminent danger fails because the new land does not provide similar economic opportunities. Resettlement of people from risky areas should carefully consider the relocation of the livelihoods and the availability of services to support the already existent economic dynamics, since these communities' economy is already extremely delicate.

Given these parallel dynamics; attachment to familiar locations and social embeddedness in local economic opportunities, it is essential that any planning that attempts to move informal settlers from high risk locations must engage the settlers in identifying and accommodating the factors that define their everyday existence.

5 Informing relocation decision making and alternative strategies

Disaster mapping and risk evaluations are common activities for researchers and government officials throughout the world. Such data produce reports that create a picture of a certain location or community. Relying on numbers and maps to determine the situation of an at-risk settlement is necessary, but may not be sufficient in or by itself. It is suggested here there should be better efforts in integrating the environmental elements that shape the behavior and identity of a given community with risk mapping and assessment.

Failure to identify such social, cultural, economic and political factors leads to poorly informed decision making and policy. It also undermines the process by which relocation through mitigation or adaptive measures may be reduced or avoided. Furthermore, as discussed above, if such strategies and policies do not comply with the realities of the community, there will likely be reasonable opposition and even social unrest during the implementation processes. *Our proposition: social, cultural, economic and political factors and interactive dynamics affect the way people decide where to live in spite of risk conditions.* The outcome of this study is expected to enrich planning and emergency management efforts in cities with at-risk settlement dilemmas. It is our objective to: identify cultural, social, economic and social factors that influence people's decision of settlement. More specifically, the plan is:

- to determine the factors and social dynamics that characterize the population of human settlements and their decision to remain despite implied or confirmed risk,
- to analyze and present information about communities living in at-risk settlements,
- to create a model to inform governmental decision making on relocation and relocation alternatives.

Multiple studies to this day have individually studied human settlement tendencies and social factors shaping communities [3, 14, 15]. But the correlation between the two has received relatively little empirical attention. This study strives to answer the question: what are the processes and factors that determine why communities remain in their settlements in spite of risk? Hewitt [18] proposes that "Geography enters all levels of human organization, the security of persons and ways of life, and our sense of safety are most intimately bound up with the place of residence and the world to which we belong". It can be inferred from this theory, that the relationship between settlement and social, cultural, economic, and political factors is a direct strong one. This study strives to find a link between these factors and the decision to settle and remain in at-risk settlements. It is expected that the study of individuals or households will paint an accurate picture of communities sharing the same space. If the above is correct a list of factors shaping settlements should be expected; these factors can be ranked by level of importance. In order to compile them, a mixed methods approach is planned with up front inductive ethnographic data gathering and analysis.

Population and sampling: this study is interested particularly in the individual who is part of a community that is living in at-risk settlements and who resists relocation. The individual can be further associated with household units in a community; by sampling households this study plans to map the community's overall relocation mindset. The study will also focus on low income communities in scattered settlements, emphasizing the developing world. The research is designed to consist of multiple case studies in which the investigators describe and identify themes and issues. The researchers will analyze data collected through interviews, observations and documents. Snowball sampling, a purposeful sampling technique, will be used to locate participants for interviews. Ultimately, this multiple case study can inform a theoretical model that seeks to reduce the loss of life in high risk areas.

6 Pilot project: Philippines

- *Barangay Banaba:* Municipality of San Mateo, Rizal Province, Philippines.
- *Population:* Banaba is one of 15 San Mateo barangays, had a 2007 population of 20,861.
- *Description:* Barangay Banaba is located in the southwest corner of San Mateo and is bounded by the Marikina and Nangka Rivers. A few of the homes in these settlements are built from strong materials, but most are built from light or salvaged materials. The rivers, which have flooded the settlements several times in recent years, pose the most significant threat to these settlers and their homes.
- *Initial findings:* the initial field work in the Philippines showed that informal communities in San Mateo often settle on public land deemed unsuitable for development because of its inherent risk; the inhabitants typically cannot afford to live elsewhere. Many have abandoned poor rural regions that lack opportunities for the appeal of distant cities. Our informants suggest that these urban poor resist relocation because their lives, however dismal, are built around informal economies near their slum settlements. Some of them describe results from prior re-settlement programs in which the settlers discover that relocation sites lack these economic opportunities, and they promptly return to the same or similar urban slums.

Under the media spotlight following disasters, government leaders often want to be seen taking definitive action. Following extensive flooding and numerous resultant drownings in Manila in August 2012, for example, public officials drew special attention to those living in creeks and along rivers because "they serve as obstructions to the easy flow of water" [23]. In this account, informal settlers are no longer vulnerable casualties of the flooding: they are its culprits. Therefore, orders came from cabinet officials to begin clearing 125,000 informal settlers and relocating them to other government designated sites. The administration initiated an embarrassing backtrack soon afterward when it was revealed that several designated relocation sites had themselves been inundated.

The authors' recent field evidence suggests, moreover, that despite good intentions, the top-down plans of government agencies are often built around simplistic monetary incentives that proved unsuccessful in the past. Having studied informal settlements in situ leads to the observation that these plans seriously underestimate the social factors – family, neighbors, informal economic networks, daily routines of family members – that guide the everyday lives of informal settler families.

7 Summary

In this, and our subsequent field studies, the intention is to identify broader and deeper relationships between human settlements and the willingness to deal with risk. The authors believe that the findings of their study and their refined instrument will be eminently useful to inform governmental post-or pre-disaster relocation or alternative strategies for mitigating the risks to informal settlements. They are also convinced that a complex array of individual, community and socio-economic factors underlie the sometimes fatal decisions of informal settlers to remain in risk-prone locations. By identifying and ordering this set of factors and the ways they interact, it is possible to inform and enrich government policy making and overall disaster risk reduction efforts. Improved policy making of this type saves lives and fosters cooperative relations between members of the community and those charged with overseeing their wellbeing.

References

[1] Innes, J.E. & Booher, D.E, *Planning with Complexity. An Introduction to Collaborative Rationality for Public Policy*, 1st ed., Routledge: New York, pp. 89-117, 2010.

[2] Fainstein, S., *The Just City*, Cornell University Press Sage House: Ithaca and London, USA, pp. 5-86, 2010.

[3] Jha, A.K., (ed). *Safer Homes, Stronger Communities: A Handbook for Reconstructing after Disasters*, Global Facility for Disaster Reduction and Recovery (GFDRR). The World Bank: Washington DC, 2010.

[4] Scott, J.C., *The Moral Economy of the Peasant: Rebellion and Subsistence in Southeast Asia*, Yale University Press: New Haven, CT, pp. 157-230, 1976.

[5] Friedmann, J., *Planning in the Public Domain: From Knowledge to Action*, Princeton University Press: Princeton New Jersey, pp. 19, 220-284, 1987.

[6] Healey, P., *Collaborative Planning: Shaping Places in Fragmented Societies*, 2nd ed., Palgrave Macmillan: New York, pp. 49-62, 2006.

[7] Scott, J., *Seeing like a State: How Certain Schemes to Improve the Human Condition have failed*, Yale University Press: New Haven, CT, pp. 214-455, 1998.

[8] Scott, J.C., *Weapons of the Weak: Everyday Forms of Peasant Resistance*, Yale University Press: New Haven, CT, pp. 16-38, 1985.

[9] Satterthwaite, D., *What Role for Low-income Communities in Urban Areas in Disaster Risk Reduction?* Global Assessment Report on Disaster Risk Reduction, UNISDR: London, pp. 1-48, 2011.

[10] Lewis, D. & Mosse, D., (eds). *Development Brokers and Translators: The Ethnography of Aid and Agencies*, Kumarian Press: Bloomfield, CT, 2006.

[11] Borgatti, S.P., Mehra, A., Brass, D.J., & Labianca, G., Network analysis in the social sciences. *Science*, **323**, pp. 892-895, 2009.

[12] Bourdieu, P., The forms of capital, *The Handbook of Theory and Research in the Sociology of Education*, ed. J.G. Richardson, Greenwood Press: New York, pp. 241-258, 1986.

[13] Forester, J., *Planning in the Face of Power*, University of California Press: Berkley, Los Angeles, London, pp. 48-64, 1989.

[14] Granovetter, M.S., Economic action and social structure: The problem of embeddedness. *American Journal of Sociology*, **91**, pp. 481-510, 1985.

[15] Schumacher, E.F., Patterns of human settlements. *Ambio*, **5**(3), pp. 91-97, 1976.

[16] Hewitt, K., (ed.), *Interpretations of Calamity: from the Viewpoint of Human Ecology*, The Risk and Hazards Series 1, Allen & Unwin Publishing: USA, pp. 20-27, 1983.

[17] Hardoy, J.E., & Satterthwaite, D., *Squatter Citizen: Life in the Urban Third World*, Earthscan Publications: London, 1989.

[18] Hewitt, K., *Regions at Risk. A Geographical Introduction to Disasters*, Addison Wesley Longman: USA, 1997.

[19] Nidhiprabha, B., *Adjustment and Recovery in Thailand Two Years after the Tsunami*, The Asian Development Bank Institute, Discussion paper No.72 :Tokyo, Japan, pp. 22-37, 2007.

[20] Oliver-Smith, A.S., *The Martyred City: Death and Rebirth in the Andes*, 2nd ed., Waveland press: Prospect Heights, IL, pp. 5-27, 1986.

[21] Bode, B., No bells to toll: *Destruction and Creation in the Andes*, 1st ed., Scribners: New York, pp. 169-237, 1990.

[22] LaFerrara, E., Self-help groups and income generation in the informal settlements of Nairobi. *Journal of African Economies*, pp. 61-89, 2002.

[23] Villas, A.L.T., Robredo: Act on informal settlers. News article: Tempo.com. August 17, 2012. Retrieved from: http://www.tempo.com.ph/2012/robredo-act-on-informal-settlers/

Hydrometeorological effects of climate change on the habitat in Southern Mexico

J. F. Audefroy
National Polytechnic Institute IPN, Mexico

Abstract

This study evaluates the potential effects of climate change on the habitat and human settlements in Mexico through an analysis of three regions that are vulnerable to hydrometeorological hazards such as droughts, floods and hurricanes. The research process included fieldwork in the states of Oaxaca, Tabasco and Yucatán, as well as a historical study of hydrometeorological events in each region. The aim was to identify a means of interpreting these events, linked to climate variability, on the basis of the history of disasters, the environment and the habitat. The local climatic indications were compared to the global successes of the Intergovernmental Panel on Climate Change (IPCC), to show that contradictions do not exist but that it is difficult to apply the IPCC's findings at a local level, given the considerable margin of uncertainty. The indications of the effects of climate change make it possible to foresee that the most vulnerable populations will be the ones facing the strongest impact in the future. The reported research has direct implications on urban and housing policies, offering a roadmap to design climate change adaptation strategies; the possibility for adaptation not only requires political commitment, but is also related to social and economic development and an "integral risk management" approach, rather than a "civil protection" strategy.
Keywords: climate change, habitat, risks, hydrometeorological hazards, Mexico.

1 Introduction

Although there is no "theory" as such on adaptation to climate changes, several theoretical positions regarding this theme have been produced within the fields of history and anthropology. Some communities have also developed adaptive strategies of a social, technical and cultural type to face environmental and climate changes. This is a way in which traditional habitats have been adapting to different climates. From the perspective of political ecology, Oliver-Smith [1] has made several contributions to the subject, stating that disasters result from a historically constructed pattern of vulnerability that is based on location, orientation, existing infrastructure and the sociopolitical conditions. Disasters are thus complex, non-lineal, multicausal and multifactorial processes. In addition, there is a climate-oriented approach to risks based on a systemic analysis produced by Pigeon [2]. He states that disasters comprise a "feedback" phenomenon, that there is an interaction between the different variables or parameters and that there is not necessarily a cause-effect relationship between the different aspects of a disaster.

In a recently published article, Pelling [3] considered that adaptation in response to any geological or hydro-meteorological event may imply material or institutional changes. In the habitat field, changes can be applied to the decision-making structures or to the construction design, materials or standards (as was the case in Mexico City following the 1985 earthquake). At a local level, the adaptation process can be conducted through families, communities and/or local authorities.

Based on the study of three different cases in Mexico located in the states of Oaxaca, Tabasco and Yucatan regarding three different phenomena – drought, floods and hurricanes – the adaptive practices followed by communities and municipalities to face the effects that climate change has on habitat shall be observed. In the first case (Tilantongo in Oaxaca State), recurring drought forced both communities and authorities to take measures to harvest and store water both for human consumption and crops. In the second case (Villahermosa in Tabasco State), the floods in 2007 forced the state to displace the affected populations to new housing settlements beyond the floodplains. In the third case (San Felipe and Cacalchen in Yucatan State), following Hurricane *Isidore* in 2002, the authorities conducted a reconstruction program comprised of basic housing units in the same plots of land affected by the hurricane. These three adaptive strategies designed to deal with climate variability posit different issues. This piece of research assessed the difference in vulnerability among families, beyond the effects of climate change as a combination of socio-economic inequalities and habitat type vis-à-vis exposure to natural threats, droughts, floods and hurricanes. Lastly, it conducted a comparative evaluation of local climate trends vis-à-vis large-scale trends identified at a global level by the Intergovernmental Panel on Climate Change (IPCC).

In view of the fact that few studies have integrated a spatial dimension into the human-climate-environment relationship, this piece of research introduced habitat as a new element in climate anthropology.

In the presence of climate variability, such as drought, floods or hurricanes, this variation impacts both community and habitat. These three elements, climate, community and habitat are physically and symbolically intertwined.

2 Background information: three case studies

La Mixteca Alta, inhabited by Mixteco people (*ñuúsavi*) and located in Oaxaca State and Tilantongo, is one of the 32 municipalities of the Nochixtlán District. Except for the municipal seat, Tilantongo is comprised of dispersed settlements formed by ranches far from the administrative center. Ninety per cent of the municipality's population is estimated to live on ranches. The climate in La Mixteca Alta is mainly dry. It is nonetheless considered to have a temperate climate. Although it rains in the summer, the humidity levels are low. During the rest of the year, there is little or no precipitation. Except for the period from 1989 to 2001, it could be said that there is drought in Oaxaca State every five years, having suffered the experience of extreme drought in 1982. The lowest annual rainfall in Nochixtlán District was 322 mm between 1926 and 1930; 427 mm between 1945 and 1948 [4] and 440 mm in 1984 (INEGI, 1984). The inhabitants' memories confirm these figures. The annual average for Oaxaca State between 1961 and 1990 was 1,349.60 mm [5].

The city of Villahermosa in Tabasco State is located in the watershed of the Grijalva and Usumacinta rivers. In pre-Hispanic times, it was a favorable place for agricultural development and a water-based civilization. The hydraulic works constructed by the Chol and Chontal peoples provide testimony of this. In the studied period (from 1652 to 2008) for which hydro-meteorological events were recorded, there was recurrent flooding particularly in September, October and November. The catastrophic flood in October 2007 was not an isolated phenomenon. Based on this information, an overflow rate of two events every ten years can be estimated for precipitation of over 500 mm and a rate of 1.33 events for precipitation over 700 mm every ten years.

In Yucatan State, hurricanes, their effects and the reconstruction program, as a result of *Isidore* in September 2002, in particular, were considered. Two case studies were analyzed. The first was carried out in Cacalchen, located 41 km east of Merida City, and the second one took place in a fishing village called San Felipe on the northern coast of Yucatan. Geographically and ecologically speaking, these two villages are very different from each other. Cacalchen is located in the old production area of *henequén*, an agave from which a fiber resembling sisal is obtained, an area in which the traditional Mayan dwelling prevailed, that is, a house made of *bajareque* (a lattice made of sticks and mud) or stones with an apsidal roof made of *huano* palm leaves. San Felipe is a fishing village in which the houses are made of wood with a gable roof made of *huano* or tin sheets. The two villages were severely affected by Hurricane *Isidore*.

3 Climate trends observed in Oaxaca, Tabasco and Yucatan states

3.1 Oaxaca State

It is interesting to observe that different sources of information evidence an alternation between drought, abundant rainfall and frosts. From a historical perspective, it can be deduced that, at least since colonial times to date, the same phenomena have appeared in succesion. These phenomena have been recurrent, although with varying alternations. In the period from 1970 to 2011, it can be observed with more precision that drought was a recurrent theme among the Oaxacan and Mixteco populations. In this same period, in accordance with a database called *desinventar (*an on line Inventory System of the effects of disasters produced by La Red for North, South and Central America), a recurrence of drought was observed (fig. 1).

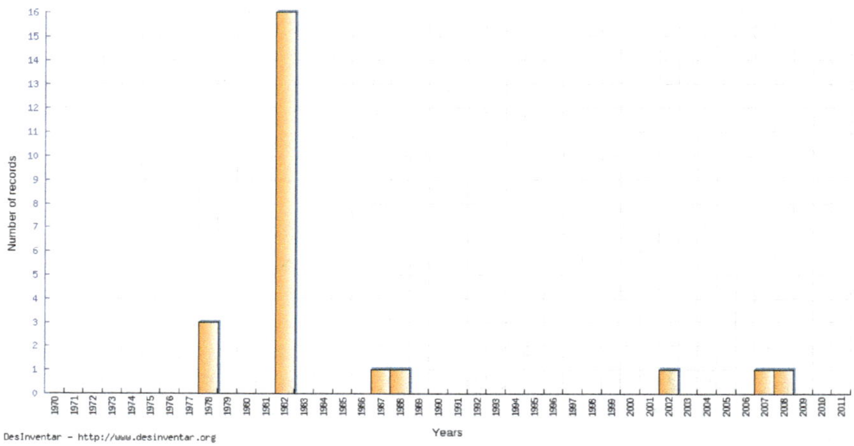

Figure 1: Drought in Oaxaca State, 1970–2011. (Source: *Desinventar*. Available at: *http://www.desinventar.org*)

During this same period, from 1997 to 2011, there were intense rainfalls. Prior to that period, intense rainfall was sporadic. It can be observed that the *El Niño* phenomenon in 1998 did cause intense rainfall in Oaxaca State, although it was less intense than that of 2010 and 2011.

3.2 Tabasco State

Historical data reveal that the risk of floods in Tabasco State is a social construction due to anthropogenic activity over time: river diversion, tree felling and the introduction of cattle-ranching have contributed towards changing the state's ecosystems and hydro-geological context. In the 1940s, unprecedented intense deforestation was noted in the region [6].

The graph in fig. 2, regarding the number of floods for the period from 1979 to 2011, was obtained from the *Desinventar* database. This figure shows that the number of floods in Tabasco State was higher since 2005 with high peaks in 2007, 2008, 2010 and 2011. This fact has led researchers to associate these floods with the effects of climate change in Tabasco State for those years. Nonetheless, these statements will be considered as an unproven hypothesis.

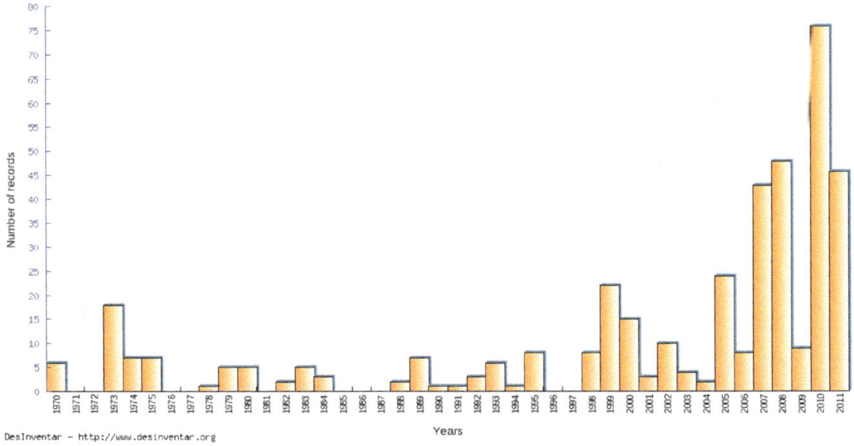

Figure 2: Number of floods in Tabasco State in 1970–2011. (Source: http://online.desinventar.org/)

3.3 Yucatan State

Of the 54 recorded hurricanes and tropical storms, reported to have affected the Yucatan Peninsula since the Conquest up to 2009, in total 13 hurricanes left an imprint. Taking into account another four hurricanes that approached the peninsula in recent times, the figure raises to a total of 17.

Thanks to *Desinventar.com* there is more reliable data for the period from 1970 to 2011. This database provides information on floods and rainstorms in Yucatan State corresponding to tropical storms and hurricanes.

Fig. 3 demonstrates that hurricanes have generated floods and rainfall that were not that frequent in earlier decades. Statistics show that in the later 14 years, i.e., from 1998 to 2011, these phenomena increase. During the period from 1995 to 2007, six high-intensity hurricanes were recorded in Yucatan. Both the rainstorm and flood graphs as well as recorded hurricanes show certain changes in hydro-meteorological phenomena in the last two decades. The higher intensity of hurricanes and the abundance of floods and rainfall seem to be pointing towards a new trend. Of course, not all hurricanes qualify as disasters. However, during the period from 1980 to 2007, twelve hurricanes in the Gulf/Caribbean could be considered disasters. The hurricane category does not necessarily coincide with its destructive impact. For example *Isidore* (Cat. IV, 2002) with a lower category than *Gilbert* (Cat. V, 1988) resulted in much more devastating effects. It is worth asking what might have happened with these hurricanes if society were less vulnerable, if the environment were less destroyed and if there were a greater recovery capacity and a more developed resilience.

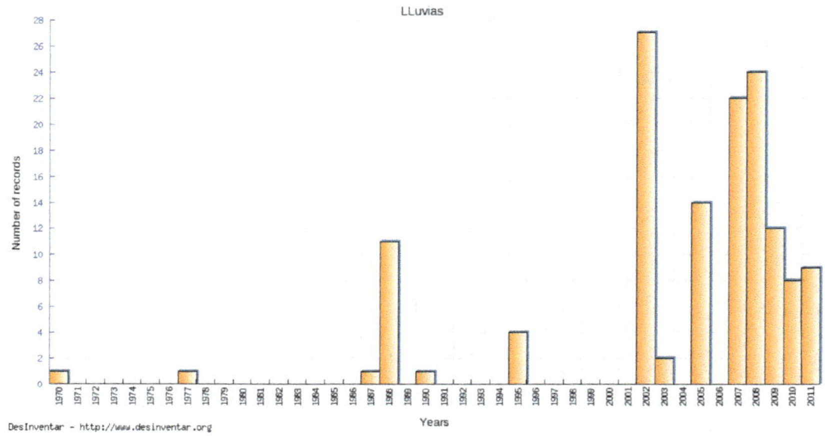

Figure 3: Rain storms in Yucatan State – 1970–2011 (Source: http://online.desinventar.org/)

4 Drought and its impact on habitat

The IPCC posits that climate change will affect "*water resources in some dry regions at mid-latitudes and in the dry tropics, due to the altering of precipitation and evapotranspiration, and in areas dependent on snow and ice melt*" [7]. What has been observed in the Mixteca Alta in Oaxaca State would seem to confirm this statement. Water availability will not only affect crops (more high-risk crops will be needed), but will also affect habitat. As has been observed in the Mixteca Alta, the habitat is scattered and water availability for home use will be affected. Many rivers are drying out, some water springs are decreasing, and there is a higher risk of uncontrollable forest fires. As far as health is concerned, there is a higher risk of mortality due to temperature-related causes, particularly among the elderly. In tropical zones located under 1500 meters above sea level the quality of life has worsened owing to temperature increase. Adequate housing is lacking. Precarious housing does not offer enough thermal steering to curb the effects of excessive heat. Houses have tin roofs and walls made of wooden planks, asphaltic cardboard or tin sheets. Traditionally constructed houses made of adobe or stone offer high thermal steering that limits heat inside the house.

Cases of drought in Mexico have been well documented: Sánchez-Mora [8]; García Acosta [9] and Florescano and Swan [10] set forth a chronology of droughts dating from the year 1500 to the twentieth century. As we have seen in the case of Oaxaca, recurrent drought has been an ongoing feature.

4.1 What are the future implications of drought?

According to Hernández Cerda *et al.* [11], of the seven processes associated with drought-produced desertification, three have a direct impact on habitat:

- *Degradation of the vegetable layer (deforestation):* in general, this is an anthropogenic process rather than a natural phenomenon. However, drought also contributes to deforestation. Deforestation implies consequences on habitat: trees protect houses from wind, encouraging a micro-climate that limits surplus heat. They are an excellent renewable resource and offer construction materials that are less polluting than cement production.
- *Water erosion*: it is the process of soil removal due to the action of water. Erosion may lead to landslides and differential settlement of housing foundations thus affecting construction. Alternate cycles of drought and rainfall particularly affect the soil, especially the topsoil.
- *Wind erosion*: the detachment and shedding of soil particles caused by wind. The shedding of particles may affect constructions, particularly those made of adobe, which are impacted by wind and air-transported rain.

Desertification usually decreases soil resistance. Thus, some soils that used to be suitable for construction become unsuitable and may generate settlements at risk. According to Hernández Cerda, "the soil and most of the dry land elements are, in general, resistant and able to recover after suffering climate alterations and man-made effects, such as deforestation. However, when land is degraded, its recuperation capacity is substantially reduced, which implies physical and socioeconomic repercussions".

Hérnandez Cerda *et al.* [11] note that "the region of Oaxaca constitutes a case in which the 1972 drought had disastrous effects on a destabilized ecological system". Environmental alterations caused by human activity and the prevalence of subsistence crops that are highly vulnerable to drought brought about serious economic loss and affected social issues.

On a local or regional level, it is difficult to identify whether drought and temperature increases are caused by climate change or deforestation. In fact, it is difficult to differentiate climate causes from others types of causes such as anthropogenic activity. According to the IPCC [7], the observations obtained in all continents evidence that many natural systems are being affected by regional climate changes, particularly by temperature increase. These alterations, however, also result from local human activity, such as deforestation and grazing in what used to be arable soil.

5 Impact of hurricanes and floods on human settlements

5.1 Why are the Mexican Caribbean coasts now more vulnerable to hurricanes?

In order to evaluate the vulnerability of the Caribbean coasts to hurricanes, coastal population growth must be taken into account. At a national level, from 1940 to 1950 the urban population inhabiting coastal locations doubled from 606 thousand to 1.2 million inhabitants [12]. In 1995, the coastal urban population increased to 12.6 million and the coastal population in the Caribbean-Gulf of Mexico region went from 1,069,357 in 1950 to 4,891,572 inhabitants in 1995 (INEGI, 1996). In other words, it quadrupled in 40 years. From 1950 to 1960, smaller locations also grew. For the fishing town of San Felipe, which was studied, out of a total of 1,789 inhabitants (2010), 1,769 inhabitants are considered vulnerable to sea level growth. It is known that during Hurricane *Isidore* in 2002, sea level rose by 1.50 m throughout the village.

5.2 Will hurricanes become more intense in the future?

If the cyclone intensity in the Yucatan Peninsula since 1980 is examined, 12 hurricanes considered "disasters", i.e., with a high level of intensity (category III, IV or V) can be observed. As noted above, during the 1970s, 1980s and early 1990s, intense hurricanes were less frequent than in earlier decades. However, since 1995, intense hurricanes have become much more frequent. This is what the IPCC projects for the future:

Based on a range of models, it is likely that future tropical cyclones (typhoons and hurricanes) will become more intense, with larger peak wind speeds and more heavy precipitation associated with ongoing increases of tropical sea-surface temperatures [7].

5.3 Will hurricanes be more numerous and intense in the future?

The IPCC [7] shows data pointing to an average heating of at least 1°C on the sea-surface, which favors the formation of hurricanes and other phenomena never before recorded, such as the first hurricane in the southern Atlantic in the Brazilian state of Santa Catarina in March 2008. The information that has been collected on the Yucatan Peninsula does not allow for a confirmation of a clear trend regarding hurricane and tropical storm intensity and frequency.

6 Adaptations, mitigations and resilience

The adaptation issue is nothing new. As García Acosta [13] underlines, "societies, throughout the world and throughout history have developed social and cultural pathways to manage risks and face real and potential disasters. Societies are not, nor have been passive subjects in the face of natural threats". The IPCC is also in accord with that idea that "societies across the world have a long record of adapting and reducing their vulnerability to the impacts of weather – and climate-related events such as floods, droughts and storms" [7].

6.1 Habitat adaptation to climate variability

The case studies in Tabasco, Yucatan and Oaxaca States have demonstrated that housing's vulnerability to flooding does not depend so much on external factors such as rainfall and storms, but rather on internal factors such as exposure, location or design. In fact, a house that is hardly exposed to flooding because of its location will not be affected by rainfall and will not be at risk.

Construction design can take into account high precipitation rates, for instance, through the use of pilings, like those used in stilt house settlements. In Tabasco State, after the flooding in 2007, the Tabasco Housing Institute (Instituto de Vivienda del Estado/INVITAB) constructed 120 stilt houses in the municipalities of Centla, Centro, Cunduacan, Jonuta and Macuspana. As a result of the floods in 2007, the same institution carried out a reconstruction program in Villahermosa in three housing complexes, but in none of them were houses constructed on stilts. None of the houses had lofts, either. In other words, no lessons were learned from the floods in 1999 and 2007. The design of low-cost housing in Tabasco continues to be the same as before the catastrophic events.

So far, the best drought adaptation processes in Oaxaca State have been conducted by the Department of Agriculture (SAGARPA by its acronym in Spanish), thus promoting roof harvesting of rainwater to be stored in tanks. In the face of a deteriorated environment, the Municipality of Tilantongo designated an ecological area of 3 thousand hectares of oak forest to be preserved and approximately 300 hectares have been reforested. However, adaptation activity is relatively scarce and the municipal seats have done nothing to harvest water or promote sustainable technologies.

As a result of Hurricane *Isidore* in 2002, in Yucatan, the FONDEN program (the National Fund for Disaster Relief), changed settlement patterns, introducing a 22 m^2 cement block core house in the middle of traditional plots with a design that is foreign to the Mayan house, ignoring the tropical weather and the local context in Yucatan. All FONDEN core houses followed the same pattern, both in floodable areas (San Felipe) located on unstable soil and in zones suitable for construction (Cacalchen).

Mexico does not have an adaptation policy or strategy to face climate variability. Even though there is a National Climate Change Strategy and Municipal Climate Action Plans at a local level (PACMUN), there is no housing policy and, in particular, no low-cost housing policy that takes hydro-meteorological threats into account. The adaptive capacity is not only related to political will power, but also has to do with social and economic development. The IPCC posits it as follows, *Adaptive capacity is intimately connected with social and economic development, but it is not evenly distributed across and within societies* [7].

7 Conclusions

The population sector is currently more vulnerable and exposed to consequences attributable to climate change. Although the business sector is also exposed (as shown in the Global Assessment Report No. 13 [14]), it has the means, e.g. insurance and government support, to recover from disasters. The effects of climate change have a direct incidence on the correlation between development and justice in the exposed populations. The gap is opening more and more as impacts and disturbances become stronger in different realms: the economy, society, equality, service accessibility, territory and security. The most developed sectors have greater access to information and greater possibilities of achieving resilience and mitigating immediate consequences. The sectors that struggle on a daily basis in order to achieve a more decent life, will be facing more difficulties to mitigate the effects of disasters. They are caught between an "assumed freedom" and the governments' authoritarian actions that condition their reconstruction programs, ignoring the specificities of vulnerable groups.

In the medium and long term, in the dawn of the twenty-second century, the more vulnerable populations will undoubtedly face a scarcity of drinking water and an increase in its cost, as well as an increase in the risks caused by landslides and floods, particularly in densely populated areas on the coast. Other populations will face drought in some areas and its immediate consequences: temperature increase, which places the whole issue of energy cost at stake. Others will also be facing more intense or longer lasting hurricanes that will endanger the family economy, which, to start with, is precarious enough.

This research also revealed that whereas urban development actors are in charge of assisting the populations affected by hydro-meteorological events, through applying reconstruction programs, they are contributing towards an increase in the risk of suffering disasters and catastrophes. In fact, this research demonstrates that the development of human settlements and ensuing inappropriate practices, are not only affected by climate change, but are also one of its main causes, since they might:

- Increase vulnerability;
- Increase exposure to extant risks;
- Intensify/amplify urban risk and create new risks;
- Lead to vulnerability and the risk of constant change (which makes them practically impossible to control);
- Reduce the capacity to face risk at a national and municipal level; and
- Reduce the capacity that low-income households and communities have to respond locally, as in the three cases.

In addition, vulnerability variables were identified to underlie the complex system of risk and disaster occurrence in poor neighborhoods, in urban neighborhoods and also in rural and semi-rural settlements. This situation of social vulnerability intensely promotes the vicious circle of existing poverty in which people are trapped.

To conclude, like the IPCC, prudence must be exercised regarding climate change: disasters are multi-causal and at the same time result from a population's high vulnerability, human action in the environment and the effects of climate change. However, this can only be demonstrated locally, as has been analyzed in the three aforementioned case studies.

Acknowledgements

The current work is a spin-off from research SIP: N° 20131120 and N° 20141159, funded by the Research and Postgraduate Department (Secretaría de Investigación y Posgrado) pertaining to the National Polytechnic Institute of Mexico (Instituto Politécnico Nacional).

References

[1] Oliver-Smith, A., Theorizing vulnerability in a globalized world: a political ecological perspective. *Mapping Vulnerability, Disasters, Development & People*, eds. G. Bankoff, G. Frerks & D. Hilhorst, Earthscan: London, pp. 10-24, 2004.

[2] Pigeon, P., Enjeux et vulnérabilités cachées: Evolutions récentes en géographie des risques. *Risques et Environnement: Recherches Interdisciplinaires sur la Vulnérabilité des Sociétés*, eds. S. Becerra & A. Peltier, L'Harmattan: Paris, pp. 53-64, 2009.

[3] Pelling, M., The vulnerability of cities to disasters and climate change: a conceptual framework. *Coping with Global Environmental Change, Disasters and Security, Hexagonal Series on Human and Environmental Security and Peace 5*, eds. H.G. Brauch, U.O. Spring, C. Mesjasz, J. Grin, P. Kameri-Mbote, B. Chourou, P. Dunay & J. Birkman, Springer: Heidelberg, pp. 549-558, 2011.

[4] Butterworth, D., *Tilantongo, Comunidad Mixteca en Transición*, INI, Mexico, 1975.

[5] Martínez Austria, P.F. & Patiño Gómez, C. (coords.), *Atlas de Vulnerabilidad Hídrica en México ante el Cambio Climático, Efectos del Cambio Climático en los Recursos Hídricos de México*, Vol. III, IMTA, Jiutepec, Morelos, 2010.

[6] Tudela, F. (coord.), *La Modernización Forzada del Trópico: el Caso de Tabasco, Proyecto Integrado del Golfo*, COLMEX-CINVESTAV-IFIAS-UNRISD, Mexico, 1989.

[7] IPCC, *Cambio climático: Informe de Síntesis. Contribución de los Grupos de Trabajo I, II y III al Cuarto Informe de Evaluación del Grupo Intergubernamental de Expertos sobre el Cambio Climático*, publication directors R.K. Pachauri & A. Reisinger, IPCC, Geneve, Switzerland, 2007.

[8] Sánchez-Mora, E., Las sequías en el México antigüo. *Análisis Histórico de las Sequías en México*, coord. J. Sancho y Cervera, Documentación de la Comisión del Plan Nacional Hidráulico, Secretaría de Agricultura y Recursos Hidráulicos, N°22, Mexico, pp. 15-20, 1980.

[9] García Acosta, V., Las sequías históricas de México. *Desastres y Sociedad*, Red de Estudios Sociales en Prevención de Desastres en América Latina, Year 1, N°1, Mexico, pp. 83-96, 1993.

[10] Florescano, E. & Swan, S., *Breve Historia de la Sequía en México*, Universidad Veracruzana, Xalapa, Ver., Mexico, 1995.

[11] Hernández Cerda, M.E., Carrasco Anaya, G. & Álfaro Sánchez, G., *Mitos y Realidades de la Sequía en México*, Textos monográficos, Vol. 6: Medio ambiente, UNAM, Instituto de Geografía, Mexico, 2007.

[12] Hernández Cerda, M.E. & Azpra Romero, E., *Los Ciclones Tropicales de México*, Textos monográficos, Vol. 6.1: Medio ambiente, UNAM, Instituto de Geografía, Mexico, 2001.

[13] García Acosta, V., Estrategias adaptativas y amenazas climáticas. *Más allá del Cambio Climatic*, compilers J. Urbina Soria & J. Martínez Fernández, INE-SEMARNAT/UNAM, Mexico, pp. 29-46, 2006.
[14] Naciones Unidas, GAR 2013, Evaluación Global sobre la Reducción de Riesgo de Desastres, UNISDR, Geneve, 2013.

The preparation and integration of Turkey's national disaster response plan

F. Oktay
Republic of Turkey Prime Ministry Disaster and Emergency Management Authority, Turkey

Abstract

Turkey frequently faces disasters that lead to human, material, economic and environmental losses, interruption of social, economic and community life, thus adversely affecting development. Located in one of the most seismically active regions in the world, 66% of Turkey's surface area lies on Zones at 1 and 2 levels of seismic hazard. Besides earthquakes, urban and rural areas need to consider other natural hazards, mainly landslides, floods, rockfalls, avalanches, that can easily combine with physical, social, economic and environmental vulnerabilities and turn into costly disasters. Under these conditions, the Prime Ministry Disaster and Emergency Management Authority (AFAD in Turkish) was established in 2009 as the authorized coordination agency to address all aspects of disaster management in Turkey, based on the reorganization of relevant major institutions as well as on the available scientific and practical knowledge, field experience of many years, resources and technology. After the Van Earthquake in October 2011, the need for developing an integrated national disaster response plan became an increasingly pressing issue on the country's agenda. New studies, approaches and plans on disaster management emerged, supported by the advancement in technology and increase in public awareness on disasters. Following detailed research on types of events and response needs, AFAD, together with governmental and non-governmental stakeholders, initiated the preparation of the national disaster response plan to define basic principles of action, activities and institutional roles and responsibilities in anticipation of, during and after disasters or emergency situations. The plan is named Turkey's National Disaster Response Plan (TAMP in Turkish) and put to effect upon official approval in January 2014. TAMP, with its flexible, modular-structure, is adaptable to all types and scales of hazards and enables effective planning of resources and coordination of public, private institutions and non-governmental organizations in case of a disaster at local or national level to minimize human and material loss.

Keywords: disaster response, integrated disaster response planning, disaster and emergency management.

1 Introduction

A disaster is a serious disruption of the functioning of a community or a society involving widespread human, material, economic or environmental losses and impacts, which exceeds the ability of the affected community or society to cope using its own resources [1]. Turkey is often faced with disasters triggered by natural hazards, which lead to human, material, economic and environmental losses because of her geological and geomorphological structure and climatic characteristics [2].

Located in one of the most seismically active regions in the world, 66% of Turkey's surface area lies on Zones at 1 and 2 levels of seismic hazard. Between 1900 and 2012 there have been 287 damaging earthquakes, leading to 100,000 deaths while nearly 700,000 buildings were reported as having major damages or totally destroyed [3]. Besides earthquakes, urban and rural areas need to be protected from other natural hazards, mainly landslides, floods, rockfalls and avalanches, that can easily combine with physical, social, economic and environmental vulnerabilities and turn into costly disasters.

After the Van earthquake in 2011, AFAD (Turkish acronym for the Prime Ministry Disaster and Emergency Management Authority) started working for a more effective disaster management policy, covering all aspects of integrated disaster management and the preparation of the National Disaster Response Plan became an effective starting step [4]. The Turkey's National Disaster Response Plan (TAMP in Turkish) covers all response action required in case of any disaster or emergency and defines the processes, activities, roles, duties, responsibilities, authorization and cooperation of all stakeholder institutions and communities at national and local levels in detail [5].

TAMP emerged as a result of a two-year study in this context. The main national policy documents searched during preparations of the plan are: Ninth Development Plan (2007–2013); Medium Term Programme (2013–2015); Rural Development Plan (2010–2013); National Earthquake Strategy and Action Plan (2012–2023); Integrated Urban Development Strategy and Action Plan (2010–2023); National Climate Change Adaptation Strategy and Action Plan (2011); Occupational Health and Safety Risk Assessment Guide.

Compared to the previous plans, TAMP was handled by adopting a comprehensive approach, the number of service groups was increased and services were thoroughly integrated. Personnel, equipment and resource analyses are to continue regularly at central and local levels, supported by up-to-date SWOT analysis to help the development and improvement of disaster scenarios, appropriate response methods and capacity in all terms.

2 The objective and principles of TAMP

The goal of TAMP is to define the roles and responsibilities of the service groups and coordination units taking part in disaster and emergency response and to determine the basic principles of a response plan in anticipation of, during and after disasters. TAMP includes personnel of the ministries, the public instutions as well as private and non-governmental organizations.

The four basic principles of TAMP are as follows:

a. Comprehensive approach (considering preparation, response, recovery phases together);
b. To address hazards of all types and scales;
c. To cover responsibilites and roles of all key and supportive institutions, partnering to provide solutions;
d. To mobilize national, regional in local disaster response capacities for fast and timely response.

The complementary basis for these basic principles, based on legislation, is defined as the establishment of an effective, flexible and scalable structure, considering improvement and further development of coordination and cooperation among all stakeholders, supported by knowledge management and efficient communications.

The objectives of TAMP are stated as follows:

- Saving lives;
- Restoring interrupted daily life and activities as soon as possible;
- Fast and planned response;
- Protect and maintain public health;
- Protect property, environment and cultural heritage;
- Reduce economic and social losses;
- Prevent secondary disasters or mitigate their possible effects;
- Ensure efficient use of resources.

3 Services

TAMP is based primarily on coordination, cooperation and collaboration. 28 service groups are established under four main service categories (namely, operation, information and planning, logistics and maintenance, finance and administration services) to ensure minimum hierarchy and maximum efficiency in implementation, using a modular structure compatible with the type and size of the event. Four main service categories are coordinated under Prime Ministry or AFAD Disaster and Emergency Management Center depending on the severity of the situation.

3.1 Operation service (Ops)

This service is the basic unit of response organization and is divided into two subservices according to required level of response and event type: for small-scale events *Emergency Services* sub-service and for large-scale disasters *Early*

Recovery Services sub-service. These services constitute the first step of the recovery process and form the basis for longer term recovery processes.

3.2 Information and planning service (I&P)

This service is to provide all services necessary for information management, monitoring and evaluation. During the response, this service is responsible for identifying the sources for needed data, data collection and analysis, conducting area reconnaissance, situation reports, situation assessment as well as for preparing alternative strategies based on current data and probability calculations and modelling by analyzing data from geographic information systems (GIS), maps, databases, remote sensors.

3.3 Logistics and maintenance service (L&M)

This service is responsible for establishing facilities, providing the necessary equipment to support all service groups involved in response activities. If the response level declared by AFAD is 2, 3 or 4, the assigned teams need to be self-sufficient and relevant institutions must be ready to meet their personnel, vehicle, equipment, office supplies and social (accommodation, nutrition, hygiene supplies, etc.) needs for "minimum 72, maximum 120 hours".

3.4 Finance and administration service (F&A)

In the scope of the preparedness for response, the financing needs should be stated in the strategic plans of public institutions within the framework of Public Financial Management and Control Law (Law no 5018). For institutions funded through general budget resources, all types of disaster and emergency management investments (met from the general budget) are monitored and coordinated by AFAD. Those institutions that have special budgets and local administrations may allocate a portion of their own budget for disaster and emergency situations. The financial management of emergency needs are made in line with Disaster and Emergency Management Presidency Organization and Duties Law (Law no 5902, article 23) and Disaster and Emergency Expenditures Directive.

4 Service groups, key and supportive partner institutions

Ensuring sustainability of interrupted services in disaster and emergency situations requires coordination and organization at ministerial level. Within this framework, national level service groups are established and for each service a public authority is assigned as a key solution provider. TAMP requires a total of 28 service groups at a national level (table 1). These service groups are established according to the needs in emergency and disaster situations. The management of these service groups are assigned to public offices and relevant NGOs (mainly Turkish Red Crescent) according to their field of work.

Table 1: National level response service groups.

Name of service group	Name of public authority/NGO	Service category
Purchase and lease	AFAD	F&A
Accounting, budgeting and financial reporting	AFAD	F&A
National and international donations (in-cash)	AFAD	F&A
Loss assessment (financial)	Ministry of Finance	F&A
Information management, evaluation and monitoring	AFAD	I&P
Service group logistics	AFAD	L&M
In-kind donations, warehouse management and distribution	Ministry of Family and Social Policies	L&M
Technical support and supply	Ministry of Transport, Maritime Affairs and Communications	L&M
Resource management	AFAD	L&M
International support and cooperation (supplies, personnel etc.)	AFAD	L&M
Search and rescue	AFAD	Ops
CBRN	AFAD	Ops
Accomodation (shelter)	AFAD	Ops
Energy	Ministry of Energy and Natural Resources	Ops
Damage assessment	Ministry of Environment and Urban Planning	Ops
Infrastructure	Ministry of Environment and Urban Planning	Ops
Debris removal	Ministry of Environment and Urban Planning	Ops
Psychosocial support	Ministryof Family and Social Policies	Ops
Food, agriculture and livestock	Ministry of Food, Agriculture and Livestock	Ops
Health	Ministry of Health	Ops
Fire	Ministry of Internal Affairs	Ops
Evacuation and placement planning	Ministry of Internal Affairs	Ops
Security and traffic	Ministry of Internal Affairs	Ops
Interment	Ministry of Internal Affairs	Ops
Nutrition	Turkish Red Crescent	Ops
Communications	Ministry of Transport, Maritime Affairs and Communications	Ops
Transportation (infrastructure)	Ministry of Transport, Maritime Affairs and Communications	Ops
Transportation	Ministry of Transport, Maritime Affairs and Communications	Ops

Key solution partners are responsible for the management of service groups and determine the roles and tasks of institutions and organizations they are to work with, the supportive partners. All service groups should work in coordination with AFAD.

The supportive partners should take part in preparation of operational plans, support the execution of operations and meet the demands of the key partners, providing trained personnel, equipment, tools etc.

After the preparation of the national service plans, the key partner (ministry, institution or organization) in charge of the service group regularly evaluates the plan and makes updates in terms of personel, supplies, equipment etc. Plans should be revised and developed based on scientific and practical knowledge and experience drawn from the field in disaster and emergency situations, national and international training and exercises as well as technological developments.

There is a total of 10 ministries and an NGO (Turkish Red Crescent) assigned as key partners and over 100 of supportive partner institutions.

5 National and local level plans and integration

TAMP serves as a binding frame for conducting response work in disaster and emergency situations, capable of adapting to all scales at local and national levels (fig. 1).

Primarily, national level service group plans are prepared together with all the national level key and supportive partner institutions under the guidance of TAMP. Thus, the roles and responsibilities of institutions and organizations at the time of disaster are to be clear for a timely, a faster and more effective response.

The next step is the preparation of provincial disaster response plans and local level service group operation plans. Through these plans, just like in the national level, the roles and responsibilities are to be clarified at the local, settlement level in case of disaster and emergency. Service group operation plans at the local level are to serve as an attachment to provincial disaster response plans. The key partner of the service group in the national level prepares the local level service group plans through their own agencies at the local or regional levels with the help of the local/regional agencies of their supportive partners. It is important to note that only two service groups are established and work at a national level: (1) National and International Donations In-Cash Service Group; (2) International Support and Cooperation Service Group (supplies and personnel etc.)

The key partner is responsible for the preparation of the national service group plans and carefully accomplishes the following tasks:

- Thoroughly identify hazards and risks that can be faced during fulfillment of duties;
- Within their service group responsibilities, identify and list all types of capacities and capabilities as strengths and resources;
- Improve the coping capacity as much as possible in order to reduce the risks and damages that can be encountered;

- Guide and implement preparations for disasters and emergency situations at all levels;
- Ensure cooperation and coordination among relevant institutions;
- Update and extend all relevant information among stakeholders;
- Organize and participate in training and exercises in their field of work.
- Establish required sub-service groups and teams;
- Organize and standardize teams' working procedures, principles and schedules at national and local levels;
- Ensure effective involvement of private sector to provide necessary resources of all types.

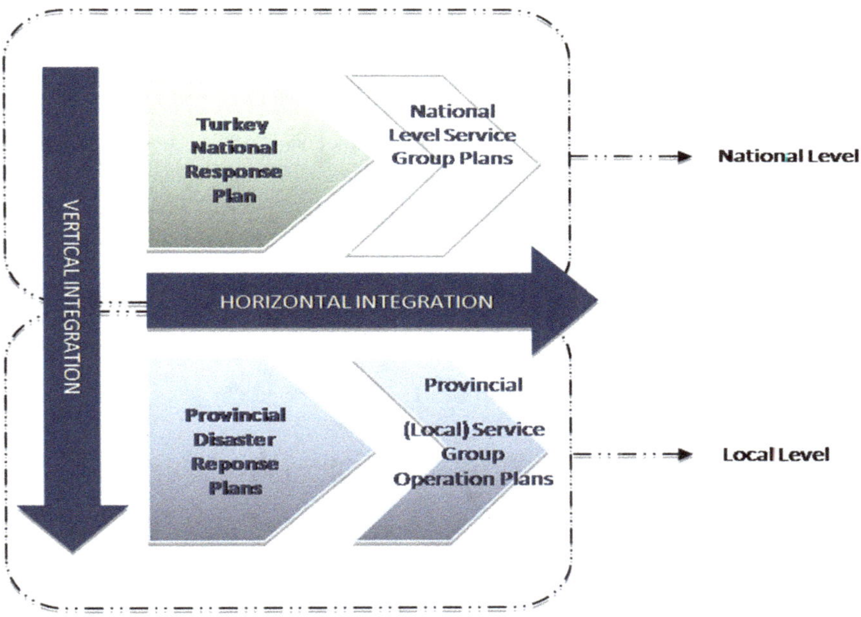

Figure 1: Vertical and horizontal integration of TAMP.

All 28 national level service group plans are prepared according to TAMP, and the groups work in the field in coordination and uninterrupted communication with Provincial Disaster and Emergency Management Centers established in Province Governorates. Each service group is informed about what other service groups may be demanding in detail beforehand. As other needs would surface in the face of disaster or emergency, the plan underlines the vitality of updating, experience sharing and capacity building, which in fact are obligations of duty.

All but two (donation and international support) national level groups plans are complemented with those of provincial level service groups, and their plan schemes are prepared at the national level. In the next step, local level authorities substantiate the plans with a special focus on hazards and vulnerability analysis of the province. The plan schemes are adaptable to respond to the needs and challenges of each province accordingly.

Every service group has field support personnel that would reach disaster zone immediately and help the response process begin through establishing necessary communication and connections between local and national levels. Depending on the severity of the situation, neighbouring provinces are to come to support the response work in disaster zone. In both cases, the personnel coming from outside the province or settlement, is informed about the system as all provinces use the same scheme, thus waste of time for adaptation and acclimation is avoided.

6 Disaster response levels and management centers

6.1 Response levels

Response levels are divided into four groups in terms of the degree of impact (table 2).

Table 2: Response levels.

LEVEL	RESPONSE	SUPPORT ACCORDING TO THE TYPE AND SCALE OF THE DISASTER/EMERGENCY
L1	Local resources are sufficient.	Province DaEMC
L2	Needs support/reinforcement from neighbour/supporting provinces	Province DaEMC – Related SAR Squad + 1. Group Supporting Provinces
L3	National level support is needed.	1. and 2. Group Supporting Provinces + National Capacity
L4	International level support is needed.	1. and 2. Group Supporting Provinces + National Capacity + International Support

Level 1: Provincial Disaster and Emergency Management Center (DaEMC) is active and AFAD DaEMC follows and evaluates the situation. If needed AFAD informs the required service groups and may use their capacity for the requested assignment.

Level 2: Provincial DaEMC is active, teams from supportive neighbouring provinces and regional Search and Rescue (SAR) brigades move to disaster zone without waiting for further orders. If needed, AFAD informs the required service groups and may use their capacity for the requested assignment.

Level 3: AFAD DaEMC is active, representatives from all service groups gather at Prime Ministry DaEMC. Teams from supporting provinces and regional Search and Rescue brigades move to disaster zone without waiting for futher orders, and national capacity is at full use.

Level 4: In addition to full national capacity used at Level 3, calls for international support are made by the government.

6.2 Disaster and emergency management centers and boards

a) *Disaster and Emergency High Council*, comprised of ministers, evaluates the information and suggestions from Disaster and Emergency Coordination Board, comprised of undersecretaries of ministries and from national and international sources, determines the principles of political directives on relevant issues and establishes binding decisions for all levels.
b) *Disaster and Emergency Coordination Board* evaluates the information in case of disaster and emergency, determines necessary treatment measures, ensures the measures are applied and supervises the whole process. The Coordination Board also provides coordination among public institutions and with private and non-governmental organizations.
c) *Prime Ministry DaEMC* is set-up on the order of the Prime Minister and under the chairmanship of AFAD President to ensure efficient response and coordination. DaEMCs are also established in key partner institutions to have uninterrupted contact and commmunication with both disaster zone and other authorities.
d) *Provincial DaEMC* works 24/7 under the chairmanship of a Deputy Governor, who is assigned by the Governor. The Provincial Disaster and Emergency Coordination Board is comprised of appointed and elected local public authorities. At times of a disaster, a high ranking AFAD official provides direct support and consultation required by the Governor in disaster zone and adequate number of field support personnel from AFAD assist local service groups. The key partner institutions also send field support personnel to disaster zone. Each service is coordinated by the deputy governor who heads the Provincial DaEMC under the supervision of Province Governor.

Thus, the national disaster response system is modelled in a triangle formed by a) Prime Ministry/AFAD DaEMC, b) DaEMCs of Key Partner Institutions, and c) Provincial DaEMCs.

7 Conclusions

The national service group plans that are developed at ministerial and institutional levels form the "Service Model" for preparation of plans, coordination and respective field implementation. Each service group is to serve as a sub-model for the system, displaying an example and guideline for local level service group plans. The national response system is designed to give qualified support to the local response system, thus the disaster zone.

The relationship between the service groups at national and local levels is a parameter on the execution of integrated response function in disaster zone. The disaster and emergency response management at local level, on the other hand, is designed as an indigenous model to ensure effective implementation and standardization, at the same time, countrywide.

Relevant international documents and work searched for field practice are: UN OCHA Coordination Tools; EU Civil Protection Mechanism; IFRC Code of Conduct; The Humanitarian Charter and Minimum Standards in Disaster Response (SPHERE Project); WHO/PAHO, Hospital Safety Index; INTOSAI, Accountability for and Audit of Disaster-related Aid.

References

[1] UNISDR Terminology on Disaster Risk Reduction, 2009.

[2] Gökçe, O., Özden, Ş. & Demir, A., Türkiye'de Afetlerin Mekansal ve İstatistiksel Dağılımı Afet Bilgileri Envanteri, Ankara, pp. 1–4, 2008.

[3] Turkey Tenth National Development Plan 2014–2018, Efficiency in Disaster Management Report by Committee of Experts, p. 4, 2014.

[4] Oktay, F., Tetik, C., Gökçe, O. & Çebi, G., New disaster management system in Turkey: A case study of Van earthquake. *The Sustainable City VIII, WIT Transactions on Ecology and the Environment*, **179**, pp. 1387-1398, 2013.

[5] Republic of Turkey Prime Ministry Disaster and Emergency Management Presidency, Turkey National Disaster Response Plan, Official Gazette, number 28871, dated 3/01/2014.

The evolution of the protection systems against natural disasters in Brazil: laws, agencies, information and knowledge

É. Soriano & W. A. M. Hoffmann
Federal University of São Carlos (UFSCar), Brazil

Abstract

Natural disasters occur all over the world in different degrees and levels, characterised by the frequency of events, the intensity of processes and the vulnerability of population groups. Brazil has always been loosely and commonly considered a safe country regarding the occurrence of natural disasters. This interpretation is due to its geographical location, which prevents the occurrence of earthquakes, volcanic activity and tropical cyclones in its territory. On the other hand, it is one of the countries that registered the highest number of natural disasters, especially those associated with geological and hydrologic processes. According to the United Nations, in 2008, Brazil was the 13th most affected country by natural disasters. The Brazilian Atlas of Natural Disasters indicates that the record of natural disaster occurrence in the country marked a positive increase by 268% in the last ten years. After the disaster in the mountainous region of Rio de Janeiro in 2011, the government sought to structure itself in order to prevent new disasters. Thus, different organizations that are part of the National System of Protection and Civil Defense (NSPCD) were created, such as: The National Center for Monitoring and Warning of Natural Disasters (CEMADEN), The System of Urban Occupation Monitoring proposed by the Ministry of Cities (MCid), The Integrated System for Information about Disasters (S2iD), and others. The aim of this study is to analyze the evolution of systems for protections and disasters risk in Brazil, as well as to evaluate knowledge gained by the departments of protection and civil defence.
Keywords: natural disasters, vulnerability, civil defence, knowledge.

1 Natural disasters

Natural disasters are caused by phenomena that occur in the biosphere and may result in a damaging event, causing death or injuries, material damage, social and economic activity interruption and environmental degradation (United Nation Office for Disaster Risk Reduction – UNISDR, 2007).

According to the UN/ISDR [1], a disaster represents a serious disturbance to a community or a society, involving human, material, economic or environmental loss on a wide scale; its impacts exceed the local community's or society's ability to cope with the damage with their own resources. The Emergency Disasters Database – EM-DAT [2] – defines a disaster as a situation or event that exceeds the local capability, demanding a request for external assistance at national or international levels.

Natural disasters occur all over the planet at different levels and degrees. This represents the apex of a continuous process that shows the imbalance between the social system forces, such as the territory occupied by society, with the natural system, through the force of a determined physical process, with the occurrence of a specific igniting process. They represent an abrupt imbalance, which generates modification and rupture in the territory functionality [3–5]. This occurs due to the combination of natural physical processes with vulnerability processes that increase risk and danger, such as: limitation of technical systems for prevention; states' omissions regarding the occupation of high-risk areas; lack of dialogue between the civil protection agents and the community; the determination of risk areas and others.

According to the National Secretariat of Civil Defence [6], natural disasters in Brazil have been recently occurring with greater intensity.

2 Natural disasters in Brazil

Brazil is located in the South American continent. It is composed of 27 Federal Units: 26 states and the Federal District. Brazil is the fifth largest country in area in the world, with 8.5 million square kilometres and it is the fifth most populous country, with a total population of 202 million inhabitants [7]. It is a country with large dimensions and complex territorial challenges, as well as high socioeconomic inequality, which results in a high number of people being vulnerable to natural processes that may cause disasters. It is estimated that there are around 500 areas at risk of landslides and another 300 are prone to flooding, with approximately five million people in a vulnerable condition.

Brazil has always been loosely and commonly considered a "safe" country, considering the occurrence of natural disasters, especially those caused by high intensity processes, which may result in a high number of deaths. This interpretation is based on Brazil's geographical location that prevents the occurrence of high intensity earthquakes, volcanic activity, extreme temperature gradients and severe tropical cyclones in its territory. On the other hand, it is one of the countries that registered the highest number of natural disasters, especially those related to geological processes, as landslides and other mass movements;

and hydrologic processes, such as river floods, flooding and drought. According to the United Nations, in 2008, Brazil was the 13th country most affected by natural disasters. The Brazilian Atlas of Natural Disasters [8] indicates that the record of natural disasters occurrence in the country marked a positive increase by 268% in the last ten years (as can be visualized in fig. 1).

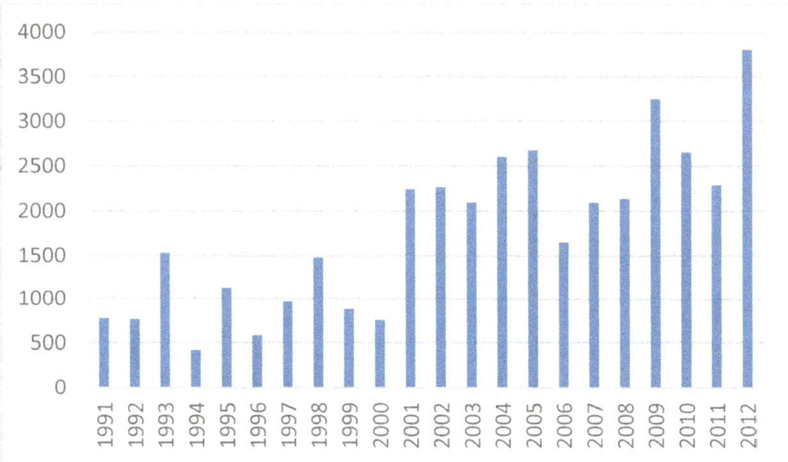

Figure 1: Evolution of total occurrence of natural disasters in Brazil [8].

According to Valencio [9], disasters are a chronic problem in Brazil and the policies adopted within the scope of the National System of Protection and Civil Defence (NSPCD) have not been able to solve this problem. In the past decade (2004–2013), Brazil presented 19,941 municipal decrees of emergency situations (ES), defined as an intense and serious deterioration of the normal situation conditions in a particular city, state or region, enacted in disaster reason, partially compromising its responsiveness; or state of public calamity (SPC), when the disaster substantially compromises responsiveness, acknowledged by the National Secretariat of Civil Defence and the Ministry of National Integration.

The urban space requires special attention for the management of natural disasters risk, since these places create or enhance risk factors. Late urbanization and the migration process from rural areas to the cities only started, mainly in developing countries, after the Second World War. This process emphasized the socioeconomic and territorial inequality of some population groups. Exclusion is mainly characterized by the access to inhabit safe areas and the construction of quality housing. In this sense, Brazilian cities may be considered risk scenes with high potential of resulting in socioeconomic and environmental damage due to: high socioeconomic inequality, lack of urban planning, absence of the state's active participation, property speculation and high population concentration, leading to a significant part of the population living in risk areas, which enhances the number of victims and affected people [10]. The risks and dangers associated with urban sprawl "are expressed by the lack of adjust and the adherence of urban space production to natural systems". The increase in these risks is not only due to natural phenomena caused by climate change, but also to urban development characterized by lack of planning, allowing the development of residential areas in areas at risk, as in floodplains and hillsides.

Brazil has a high rate of urbanization with approximately 85% of the population living in cities (as can be seen in fig. 2). According to the State Report of Latin American Cities and the Caribbean [12], in 2020, approximately 90% of the population of Brazil, Argentina, Chile, Paraguay and Uruguay will be living in cities. Nowadays, the Latin American and the Caribbean regions are considered the most urbanized regions in the world with an approximate rate of 80%. The highlights are the cities of Sao Paulo and Rio de Janeiro, combined with the cities that are part of the metropolitan region, resulting in an urban agglomeration of 21 and 12 million people, respectively.

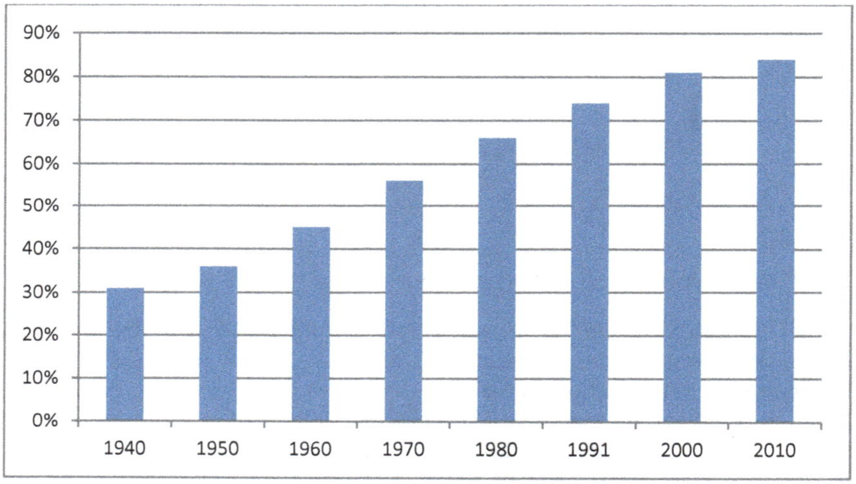

Figure 2: Evolution of urbanization rate in Brazil (source: Demographic Census IBGE).

According to Gonçalves and Sousa [13], Brazil presents a paradox according to which the economic growth is not reflected in its development, that is, it did not benefit the entire population. This process can be observed in the social vulnerabilities (poverty, low educational level and precarious access to health and housing) of large Brazilian cities, which results in the occupation of areas environmentally vulnerable (floodplains, hills and contaminated places) and irregular ways, without basic infrastructure and a building pattern that would decrease the risks.

These situations are common in almost the entire national territory. However, the risk increases with disorganized occupation of coastal areas, which are tropical wet areas that combine frequent processes of mass movement, shallow soil, great intensity precipitation and high level of human occupation. About

90% of the total population is concentrated in such coastal areas, especially around Sao Paulo and Rio de Janeiro.

Thus, population groups of low purchasing power, in most cases, are obliged to live in dangerous environmental areas, as a consequence of the real estate market's intervention, limiting their access to housing. Based on this, it is possible to state that disasters, mainly regarding landslides and flooding in specific urban areas, are socially built [14].

In an attempt to reduce the risk of disasters, mainly the number of victims, Brazil, although belatedly, introduces an evolution in the systems of protection and population defence.

3 Evolution in the systems of protection and population defence

The Organizations of Protection and Civil Defence in the country are responsible for preventive as well as assistance and recovering actions with the aim of avoiding or minimizing the consequences of disasters.

The Brazilian Civil Defence agency was created in the 1940s. The proposal of considering Civil Defence as a strategic institution to reduce the risk of disasters occurred with the organization of the National System of Civil Defence according to the decree no. 97.274, of 12/16/1988 [15]. This decree was revoked by the decree no. 895 of 08/16/1993, which instituted the National System of Civil Defence (NSCD) in the country. According to Article 3, subparagraph I, civil defence is "the set of preventive, aid, assistance and recovering actions designed to avoid or minimize disasters, preserve the populations' moral and re-establish social normality".

Considering the natural disasters legislation, The National Policy of Protection and Civil Defence (NPPCD) was approved and sanctioned, Law 12.608, of April 10th, 2012. It involves the actions of prevention, mitigation, preparation, response and recovery designed for protection and civil defence. The NPPCD, in its second article, states as a duty of the Union, States, Federal District and Cities to adopt actions necessary to disasters risk reduction, removing uncertainty as an excuse for inaction [13].

According to this policy, the cities with natural disasters risk areas must develop:

I. Mapping of areas susceptible to landslides, flooding or geological or hydrologic processes;

II. A contingency plan of protection and civil defence and institute municipal agencies of civil defence, according to the procedures established by the central agency of the National System of Protection and Civil Defence;

III. A plan of work and services implementation to reduce the risk of disasters;

IV. Control and supervision mechanisms to avoid construction in areas susceptible to landslides of great impact, flooding or geological or hydrologic correlated processes.

In this context and with the aim of avoiding socioeconomic, environmental and human damages, the Brazilian government created in 2011 the National Centre for Monitoring and Warning of Natural Disasters (NCMWND); its activities began in December of the same year. It was a response to the worst natural disaster recorded in the country: the disaster in the mountainous region of Rio de Janeiro, in January of 2011. In that period, landslides and severe flooding caused the death of about 900 people [9, 16]. Other actions taken by the government were to hire new employees and expand the infrastructure of the National Centre for Risk and Disasters Management (NCRDM), which was connected to the National Secretariat of Civil Defence (NSCD) [9].

Figure 3: Landslides in Teresópolis (RJ) in the year of 2011 (source: [17]).

The NCMWND has the aim of developing, testing and implementing a system to predict the natural disasters occurrence in susceptible areas in Brazil. The Centre not only helps the preventive actions, but also makes it possible to identify vulnerabilities in soil use and occupation, highlighting urban planning and the installation of infrastructure. The line of priority action has the goal of providing data about risk areas related to hillside landslides and flood in the country. The Centre's main strategy is to develop partnerships with state and federal institutions all over Brazil in order to improve dissemination of information and detailed analysis to solve problems [18].

The NCRDM is responsible for consolidating the information about risks in the country, such as maps of landslide and flood risk areas, besides the data related to the occurrence of technological and natural disasters and the associated damages. By managing this information, the Centre can support states and cities in the actions of preparing the most vulnerable communities for disasters. It is the Centre's responsibility to issue and broadcast the warning of national disasters risk to the competent protection agencies from cities in the entire country [19].

It is important to highlight that there are many local and regional centres of monitoring and warning in many Brazilian cities and states. They are connected to municipal and state government and directly subsidize the local systems of protection and civil defence.

4 Information and knowledge

Although the country presents an increase in the number of institutions and a new legislation regarding the theme of disaster, the information about risk and prevention has still not reached the entire population. The simulation and training situations with civil defence are extremely individual cases.

Considering school-age children, the topic "General Notions of Civil Defence and Risk Perception" began to be inserted in the teaching units as a transversal theme of the National Curricular Parameters, which may represent the possibility of universalizing the subject and the potential of reaching all students of elementary and high schools in the country.

With regard to the scientific knowledge related to the theme, meaningful evolution in the academic environment is highlighted, considering the development of research directions, research groups as well as centres of studies and research on disasters in many states through the cooperation between states, civil defence and universities, with focus on areas of teaching, research and development related to reducing the risk of disasters. Similarly, the scientific branch of The National Centre for Monitoring and Warning of Natural Disasters has been contributing to knowledge through the development of scientific research on the theme of risks and natural disasters, in the areas of meteorology, hydrological resources, geology and natural disasters, focusing on prediction, preparation, prevention and mitigation of impact of external events.

However, mechanisms that allow dialogue between groups of population and public managers are still needed. This communication must be based on adequate information and knowledge sharing, as well as the appreciation of the population's tacit knowledge, which could contribute to identify new risks and understand the causes of vulnerability processes.

5 Conclusion

Natural disasters as the result of physical processes that reform landscapes cannot be avoided. What must be done is to adopt preventive measures that contribute to the reduction of vulnerability and prevention of vulnerability processes, resulting in damage and victim reduction, especially fatalities.

In Brazil, despite being introduced late, it is possible to observe an evolution in the systems of civil protection and natural disasters management, both at prediction and operational levels, as well as in knowledge production, in sharing information of risk and in the specific legislation. In addition, the creation and appreciation of monitoring systems and technologies of risk prevention have shown to be relatively effective.

However, it is necessary to point out that there is much to be done. The main deficiencies that need to be addressed are: lack of precise information on risks and dangers, as well as the vulnerability of population groups; local and regional differences in the protection and civil defence systems; procedures and protocols standardization; improvement in information and communication flow between institutions; lack of preventive measures and appropriate information for population groups at risk; lack of a risk culture; improvement in construction patterns, associated with reducing the vulnerability process; more efficiency from the civil defence agents; the need to follow the specific legislation related to natural disasters, such as Law 12.608, and others.

The Brazilian general consideration regarding the dynamics of Civil Defence and disasters combat consists, basically, of actions during or after the disastrous event; that is, it focuses on the mitigation of consequences and not on prevention. There is still the predominance of crisis management, based on facing a disastrous situation, rather than risk management that focuses on prevention.

Therefore, there is a high demand of structural and non-structural measures, with preventive actions, not only mitigating ones. Actions that allow the decrease of vulnerability and, as a consequence, of fatalities are needed. Thus, it is necessary to establish better articulation between the actors that compose the system of protection and civil defence, as well as a more effective participation of these institutions.

References

[1] UN/ISDR – International Strategy for Disaster Reduction. Terminology on Disaster Risk Reduction, UN/ISDR: Geneva, Switzerland, 2009.

[2] EM-DAT. The OFDA/CRED International Disaster Database, www.em-dat.net, last accessed January 2016.

[3] Nunes, L.H., Compreensões e ações frente aos padrões espaciais e temporais de riscos de desastres. *Territorium*, **16**, pp.179-189, 2009.

[4] Valencio, N.F.L. da S., Vivência de um desastre: uma análise sociológica das dimensões políticas e psicossociais envolvidas no colapso de barragens. (Chapter 14). *Sociologia dos Desastres: construção, interfaces e perspectivas no Brasil*, eds. N.F.L. da Valencio, M. Siena, V. Marchezini & J.C. Gonçalves, RiMa: São Carlos, pp. 176-196, 2009.

[5] Quarantelli, E.L. *What is a Disaster? Perspectives on the Question*, Routledge: London, 1998.

[6] Secretaria Nacional de Proteção e Defesa Civil. www.integracao.gov.br /sedec, last accessed January 2016.

[7] IBGE, http://www.ibge.gov.br, last accessed January 2016.

[8] CEPED/UFSC, Atlas Brasileiro de Desastres Naturais 1991 to 2010: volume Brasil. Centro Universitário de Estudos e Pesquisas sobre Desastres. Universidade Federal de Santa Catarina. Florianópolis: CEPED/UFSC, 2012.

[9] Valencio, N. Desastres: tecnicismo e sofrimento social. *Ciência e Saúde Coletiva,* **19(9)**, pp. 3631-3644, 2014.

[10] Saito, S., Soriano, É. & Londe, L. de R., Desastres Naturais (Chapter 1). Sensoriamento Remoto e Desastres Naturais. Sausen, T. Oficina de Texto: São Paulo, 2015.

[11] Marandola, E., Marques, C., de Paula, L.T. & Cassaneli, L.B., Crescimento urbano e áreas de risco no litoral norte de São Paulo. *Revista Brasileira de Estudos Populacionais*, **30(1)**, pp. 35-56, 2013.

[12] ONU-HABITAT. Estado de las Ciudades de América Latina y el Caribe. Rumbo a una nueva transición urbana. Programa de las Naciones Unidas para los Asentamientos Humanos, ONU-Habitat: Nairobi, 2012.

[13] Gonçalves, J.C. & Sousa, R.D., Desafios de um Planejamento integrado para Prevenção de Desastres (Chapter 8). *Segurança humana no contexto dos desastres*, eds. R.L. do Carmo & N.F.L. da S. Valencio, RiMa: São Carlos, pp. 127-140, 2014.

[14] Carmo, R.L. do, Urbanização e Desastres: Desafios para a Segurança Humana no Brasil (Chapter 1). *Segurança humana no contexto dos desastres*, eds. R.L. do Carmo & N.F.L. da S. Valencio, RiMa: São Carlos, pp. 1-14, 2014.

[15] BRASIL. Ministério da Integração Nacional. Secretaria Nacional de Defesa Civil. Banco de dados e registros de desastres: sistema integrado de informações sobre desastres – S2ID, http://s2id.integracao.gov.br

[16] Soler, L.S., Gregorio, L.T., Leal, P.V., Gonçalves, D., Londe, L.R., Soriano, E., Cardoso, J., Coutinho, M., Bacelar, L. & Saito, S., Challenges and perspectives of innovative digital ecosystems designed to monitor and warn natural disasters in Brazil. *Proc. of the International Conference on Management of Computational and Collective Intelligence in Digital EcoSystems (MEDES)*, 2013.

[17] Lacerda, A/EFE/VEJA). Revista Veja. www.veja.abril.com.br/noticia/brasil/so-16-dos-municipios-tem-plano-deprevencao-a-desastres-naturais-no-brasil

[18] CEMADEN, www.cemaden.gov.br, last accessed January 2016.

[19] CENAD, www.defesacivil.gov.br/cenad/index.asp, last accessed January 2016.

Towards an integrated early warning system and disaster management in the United Arab Emirates

A. Al Hmoudi & Z. Aziz
Salford University, UK

Abstract

Early warning systems (EWSs) provide resilience to natural hazards; they also protect economic assets and development gains. The first objective of this paper is to identify and study seven countries highlighted by the United Nations as effective, with a detailed focus on Japan, the United States and Germany, in order to develop an early warning system for the United Arab Emirates (UAE) and benefit from the experiences of those countries. The research takes into account the culture of UAE society and the multiplicity of communities, as well as the socio-economic and political environment, which can emerge from the experience of the researcher in the environment in which they live. In order to ensure the effectiveness of the system before it is applied, the study examines the current situation of the early warning system in the UAE, focusing on the level of cooperation and coordination between the institutions in the UAE community and impacts of future natural disasters. Particular attention is given to the role of the National Centre for Meteorology and Seismology (NCMS) which is networked with similar centres in the region for the sharing of information and intervention at times of need. This centre also provides meteorology monitoring with advice and warning on storms. However, members of the community are uninformed regarding the procedures carried out in the event of the arrival of early warning signals. There are shortcomings in the training process and scenarios with individuals and institutions of civil society. Natural disaster management is not only about shelters and distribution of blankets, medicines and evacuation operations. It is also about education, awareness generation, mitigation strategies, community participation and lifestyles.
Keywords: early warning, multi-hazards, practices, preparedness, risk knowledge, community, disaster management, communication.

1 Introduction

In recent years, many fatalities in disasters have been attributed to the lack of an effective early warning system (EWS). Natural disasters take on a whole other meaning when living in a developing country [1]. For instance, in Sri Lanka over 34,000 people lost their lives in the 2004 tsunami. Research has identified this as due, at least in part, to the lack of a tsunami early warning system [2]. Furthermore, the disaster caused by Hurricane Katrina in 2005 serves as one of the best examples of information available for preventing and limiting damage from hazards being useless if that information is not correctly applied and broadly shared. This latter disaster was not caused merely by natural hazards, but also by institutional failure, lack of coordination between authorities and ineffective communication among all stakeholders, including the people exposed [3]. An EWS is designed to reduce the impact of a hazardous event and, if effective, can substantially increase the numbers of survivors. An example of how beneficial an EWS can be is outlined in a review of the fatalities on the east coast of India from cyclones [4]. Following a major cyclone in 1977, over 20,000 deaths occurred, resulting in the development of an EWS including meteorological radars and emergency plans, so that when the same area was hit by similar cyclones in 1996 and 2005, the death tolls were just 100 and 27, respectively [5]. It is therefore no surprise that in 2005, EWSs were characterized as the key to 'identify, assess and monitor disaster risks and enhance early warning' in the UN Hyogo Framework [6]. In March 2005, the United Nations International Strategy for Disaster Reduction (UNISDR) conducted a global survey in over 23 countries with 20 international agencies to identify existing capacities and gaps in EWS with the intention of providing a wake-up call to governments and other agencies about the role of early warnings in reducing human and economic loss from natural hazards [7]. To be effective, the report suggests that an EWS must be people-centred (i.e. community based) and should be composed of the following four elements (or subsystems):

1. Knowledge of the risks faced.
2. Technical monitoring and warning service.
3. Dissemination and communication of meaningful warnings to those at risk.
4. Response capability [5].

UNISDR identified the use of custom-made caution messages for all inclusive groups as one of the four vital elements of early warning systems. This element however is seldom accounted for by hazard observation and EWS. In association with climate change-related vulnerability and risk assessment, the assessment of the risks should encompass both climate scenarios and a combination of hazard and vulnerability scenarios. This is a move to provide a solid decision making set up [8]. As highlighted by Renn and Walker [9], risk communication is considered a key element by the Integrated Risk Governance Structure in relation to risk control. Effective risk communication should serve the purpose to educate stake holders and the society in general on the basis of risk and equip them with ability to make proper decisions in relation to risk. Effective communication between all the stake holders involved is considered a vital mechanism in approaching risk and should, in this relation, improve awareness and effectual approach to risks [10]. According to Grasso *et al.* [11], the UN certainty of data should be efficiently conversed to the users. In addition, the people at risk should be timely warned on UN certainties [4, 8, 10]. The degree of consideration given to natural hazards such as floods, tropical storms and weather related hazards such as heavy precipitation and heat waves in comparison to slowly creeping hazard processes, for instance, salinisation and rises in sea-level, the main consideration lies on the natural hazards. In addition even lesser consideration has been given to accumulated shocks from non-extreme events [8]. The major challenge in improving the ability of societies to manage environmental shocks and creeping environmental changes, is to craft an integrated structure that takes into account all relevant hazards without sidelining UN connected systems. This all inclusive approach should offer synergies and cost efficiencies, for instance in data gathering and processing, and in public preparedness endeavors. Additionally, it should aid in the enhancing and upholding warning abilities for the less frequent hazards, such as tsunamis [12].

EWS within the UAE faces various socio-technical challenges: Firstly, the role of communities within EWS is not clearly defined. Also, members of the community are uninformed and unaware of the federal plan and, as a result, they do not know the procedures carried out in the event of the arrival of early warning signals. There are also shortcomings in the training process and scenarios with individuals and institutions of civil society [13]. Moreover, there are deficiencies in the use of technology in the exchange of information and it relies on the National Centre for Meteorology and Seismology (NCMS) monitoring and warning.

Al Ghasyah Dhanhani [22] has highlighted the lack of an end to end and people-centered approach to EWS, low levels of preparedness and a general lack of knowledge regarding the emergency management plans. Secondly, the UAE has a huge expatriate population who do not speak Arabic. The presence of other languages, such as English, Persian, Hindi and Urdu create significant communication challenges. As identified by Al Ameri [23], there are significant weaknesses in the transfer of understandable warning messages and preparedness information to those at risk and there are no networking and communication among the NCMS and other stakeholders.

2 Literature review on early warning systems for natural disasters

In this contribution, natural disasters will be discussed within the environment of the United Arab Emirates (UAE) and by assessing the existing plan for natural disasters, which requires initially the identification of the kinds of natural disasters and how they should be appropriately addressed. There are no natural disasters, just natural hazards; disasters occur when people are in harm's way and preparation is insufficient. It can be seen that all individuals and communities are, to varying degrees, vulnerable to hazards and all have intrinsic capacities to reduce their vulnerability. Over the last decade, disasters have affected approximately 2.5 billion people and claimed the lives of nearly 900,000 [14]. Frequent and catastrophic disasters have increased the role of the public sector in managing disasters and emergencies [15] and the major responsibility of managing disasters and emergencies, including informing and alerting the public, rests with local government [16]. For example, in the December 26, 2004 earthquake and tsunami that killed over 200,000 people and caused billions of dollars of damage in 12 countries around the Indian Ocean [7], many lives could have been saved if there had been an effective tsunami warning system in this part of the world [17]. Early warning systems are an important strategy to save lives. However, the magnitude of the task of designing, implementing, and sustaining early warning systems in communities is enormous [18].

3 International best practices in multi-hazard early warning system

3.1 The warning management system of the Deutscher Wetterdienst

The early warning management in Germany, the Deutscher Wetterdienst (DWD) is designed as a three-tier time-orientated warning system which defines 5 levels of risks. There are specific legal provisions for the DWD mandating warnings definitions, structures, communication, co-operation and resourcing [19]. Disaster management authorities across all states in Germany assume responsibility for the management of the system. Communication is underpinned by a commitment to a single-voice principle where warnings originate solely from the country's National Meteorological Service and are legitimised for disaster management responses [19].

3.2 Multi-hazard early warning system of the United States National Weather Service

The United States created a multi-hazard national early warning system in response to a range of natural hazards including tornadoes, flooding, storms, hurricanes, drought and tsunamis. Initial steps comprised the establishment of a National Response Framework, an Incident Command System and an Emergency Alert System functional across multiple channels. Policies were instituted to reduce risk, create hazard awareness and emergency preparedness at federal, state and local level [20]. The framework established coordinating structures, which were adaptable, flexible and scalable to ensure nationwide alignment of roles and responsibilities at all levels of government and non-government organizations and the private sector. Extensive legal mandate was incorporated in statutes, regulations, executive orders and directives [20].

3.3 Multi-hazard early warning system in Japan

Japan's vulnerability to volcanoes and earthquakes as well as meteorological events such as typhoons and heavy snow has led to extensive experience of a range of natural disasters [21]. As a result, a disaster prevention scheme there is focused on early protection against the destructive effects of uncontrollable severe events [21]. The scheme relies on advanced meteorological forecast technologies and an extensive communications infrastructure to transmit disaster prevention information. A central pillar comprises enhanced observation systems, which serve to identify precisely real-time disaster risks and provide early warnings. This in turn expedites prompt evacuation and triggers response actions from disaster management agencies, which diminish the damage from calamities [21]).

4 Current status of early warning in UAE

In the UAE, the government takes great interest in preparing for natural disasters and spends a large amount of money in order to protect lives and property. They try to keep this in parallel with economic growth and urbanization and other technological developments witnessed by the Emirates. According to the four elements of early warning (Risk Knowledge, Monitoring and Warning Service, Dissemination and Communication, and Response Capability) [7], the application of those elements to the UAE is as follows:

4.1 Risk knowledge

The Ministry of the Interior is responsible for disaster management in the country through the Civil Defence, as stated in the Federal Plan for Disaster; the Civil Defence is an integral part of the Ministry of the Interior and is under the responsibility of the ministry [22]. As the UAE is exposed because of its geographical location and owing to the number of previous geological disasters, an investigator can establish that the natural disasters occurring in the United Arab Emirates are of four varieties (see fig. 1).

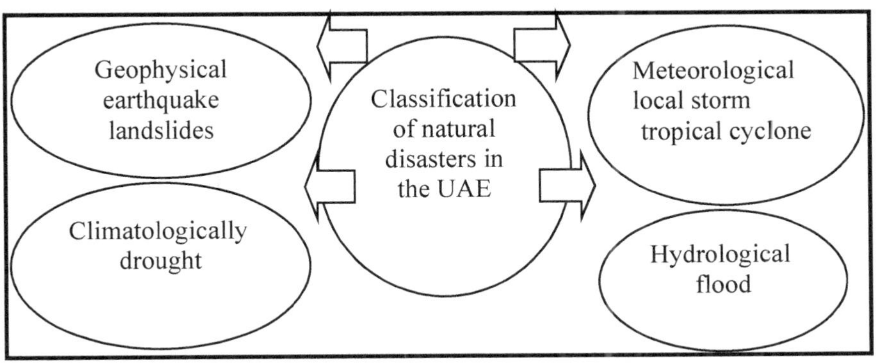

Figure 1: Classification of natural disasters in the UAE.

4.2 Monitoring and Warning Service

In the UAE and at the federal level, the body responsible for monitoring natural disasters and issuing their own warnings is the National Centre of Meteorology and Seismology (NCMS). The federal decree number 6 of 2007 was promulgated to establish and regulate NCMS [24]. The NCMS carries out its responsibilities in the framework of the World Meteorological Organization (WMO) activities and through regional and worldwide cooperation with the WMO members.

The center is affiliated to ministry of presidential affairs which is the authority concerned with meteorology and seismology affairs. It represents the country as a member in the world organization for meteorology and seismology. The center provides services to the public, governmental bodies, research centers, media, and aviation and decision makers. It performs its tasks through the following organizations and capacities:

1. Weather networks: the center operates and maintains 59 automatic weather stations all over the world, 5 weather radars and an upper atmosphere layers station.

2. Communication network: all meteorological information is to be collected and exchanged through local, regional and international communication networks. The Center is connected with the following main circles: Abu Dhabi/Muscat, Abu Dhabi/Jeddah, satellite circles and uses a Mobile Switching Station (MSS) system to collect and redistribute data.

3. Weather forecasting includes: main forecasting center, numeral forecasting center, in which a high resolution model (HRM) and its applications (SWAM – MAM) are used while Visual Weather is used to follow weather changes, flight forecasting and naval forecasting centers.

4. Rain-making applications: the center has two flights working in rain-making project and assisted by 5 earth radars. Many good experiments have been conducted in this field. This field is one of the main activities of the center.

5. Climatic studies: the center reviews and processes climatic data and prepares climatic studies and reports serving activities in different countries, especially those concerned with studies and researches such as universities, research institutions and entities concerned with applied sides and establishing a climatic information bank.

6. In an earthquake's field: the centre, through a network of five stations for monitoring strong and weak movement of earthquakes, conducts monitoring works. Such stations are the same as those stations used in earthquakes monitoring network for nuclear experiments comprehensive banned area. Also, the center is specialized in earthquake engineering studies, applied geology and the effect of soil on earthquakes waves.

4.2.1 Research, developing and training fields

The center is to perform scientific research, especially applied, in forecasting and seismology; to participate in university, research institute and related studies; to train the needed technical cadres to work in forecasting and seismology.

4.2.2 Technical services field

A department was established in the centre for delivering technical services for managing databases and systems; creating, developing and analyzing databases programming; managing and maintaining communication lines and networks; maintaining forecasting and earthquakes equipment and devices. The centre should work to provide all the developed devices and equipment for every section coping with the new developed technologies.

4.3 Challenges facing dissemination and communications in the UAE

Of the biggest challenges facing communication is the language factor, where Arabic is the official language, but the presence of other languages such as English, Persian, Hindi and Urdu and the existence of different cultures and nationalities brings economic benefits, but at the same time causes problems during disasters. The UAE is a diversified country and its native citizens only constitute 20% of the total population [23]. The geographical nature of the UAE plays an important role in the difficulty of connecting information; where each emirate has its own character in terms of terrain, such as mountains and hills; the Principality of Abu Dhabi does not contain mountains while the Emirate of Fujairah is quite mountainous, and herein lies the difficulty in warning the population. In addition, there is a large distance between the two emirates.

4.4 Response capability

It is essential that communities understand their risks in respect of the warning service and knowing how to react. Education and preparedness programmers play a key role and it is also essential that disaster management plans are in place and are well practiced and tested [7]. In addition, there is poor coordination between local partners, where it is believed that some of them face the risk of natural disasters and early warning and this is the work of the police and civil defence only. Also, members of the local community are not prepared and the reason is simple: lack of access warning to them in a timely manner and that evidence of a lack of early warning system in the state, as well as weaknesses exist in educating community members on how to behave in the event of the arrival of the warning.

5 Methodology

According to the four elements of an early warning system explained previously, the second stage is disaster observation. NCMS plays a vital role with respect to observation, weather forecast and prediction process, which is the essence of early warning system. This research has focused on explaining and studying NCMS in Abu Dhabi which is under the ministry of presidential affairs. A random sample was taken from the centre's staff to determine how staff understands their role and needs of early warning system and the availability modern technology means to early warning system and natural disaster observation.

5.1 Survey results

The study was conducted on a sample consisting of 40 persons from workers in NCMS in UAE (table 1). The research followed certified scientific approaches and frameworks in selecting the sample for study. The results are as follows:

Table 1: Sample age groups.

Age group	20–30	31–40	41–50	51–65
Percentage	(14%)	(29%)	(14%)	(43%)

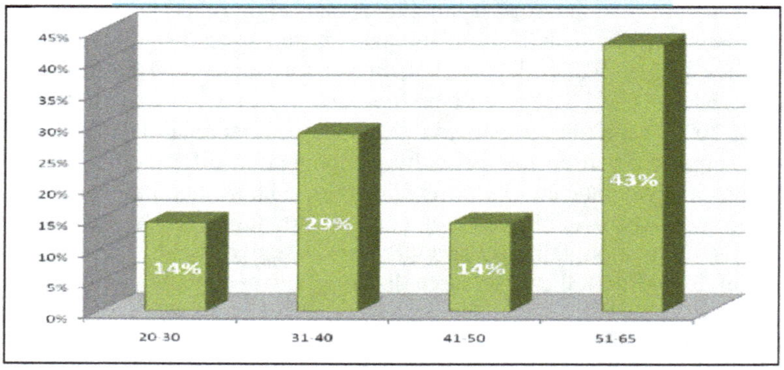

The study showed that 6 persons (86%) from the sample of study are satisfied with current weather forecast pattern performance and efficiency. On the other hand, only one person, (14%) from the sample is not (table 2).

Table 2: Satisfaction with current weather forecast pattern performance and efficiency.

Satisfaction average	Yes	No
Percentage	(86%)	(14%)

The study showed that 6 persons (86%) of the studied sample are satisfied with the current tsunami forecast pattern performance and efficiency. While only one person (14%) is not (table 3).

Table 3: Satisfaction with current tsunami forecast pattern performance and efficiency.

Satisfaction average	Yes	No
Percentage	(86%)	(14%)

The research found after a visit to the NCMS, that the best way of coordination between the Centre and the Department of Emergency and Public Safety, is by exchanging field visits to become familiar with the work nature of each other as well as exchanging data and experience. It is here suggested that a deputy from the Centre be appointed in the Central Operation room of the General Directorate of Abu Dhabi Police to provide the required information about weather updates and to explain its details, as the specialist on weather conditions observation.

6 Challenges of EWS pertaining to new environmental shocks in the UAE

The policy lessons that can be emulated from existing EWS and monitoring in the context of dissimilar and exposed target groups in NCMS (UAE) are summarized as follows.

- So as to prop up resilience building, the focus on EWS should not only be on hazards but also on the notion of exposure. There is need for EWS to be improved so that people are not only warned before the event of a hazard impacting a society, but also have a say in making sure that the livelihoods and lives at risk are secure from the gradual changes brought about by climate changes.
- There is need for information on hazard pattern changes to be merged with knowledge and forecasts about the development of vulnerabilities. The combination of the two will present the foundation of high quality EWS.
- There is no sustainability in isolated or individual hazard, especially in the context of dealing with the challenges related to climate change. This calls for the need to give an explanation for rapid and gradual hazards simultaneously; for example, the potential significance of storm surges and sea-level rises.
- Top-down, technocratic, expert-driven and linear EWSs alone are not effective at reducing damage and saving lives. A mixed institutional approach that includes both formal and informal phases is frequently indispensable. In such a scenario, the system does not control the people at risk but the people are an essential part of the system, thus creating a people-centered approach. This entails combining different types of knowledge in EWS, especially by connecting scientific and traditional knowledge.
- The development process of EWS entails a multi-disciplinary approach. This approach must recognize all the stakeholders in the community and along the chain. It must also strive to build partnership and dialogue instead of being based on a top-down alert process.
- The human assets and preparedness approaches connected to a EWS are especially imperative in low-technology places.

- There is supposed to be a change towards working, relating and communicating 'with' the community as opposed to 'for' the community.
- The shift of EWS technologies in dissimilar societal and cultural settings raises the question of robustness and societies' abilities to uptake such complicated technologies devoid of dealing with the digital divide governance problem.
- In addition, the new technologies and communication platforms like Facebook and Twitter and the internet generally have to be understood better. This is in terms of their role in enlightening and warning people in situations of calamity.
- The effectiveness and sustainability of a EWS is driven by the systems of governance, which is economic, political and socio-technological.
- The founding and development of EWS should be an essential objective of both rural and urban planning and connected sectors.
- So far, EWS have been very effective in identifying sudden-onset hazards at the same time creeping societal tipping points and environmental changes that are imperative dynamics in decisions to migrate. In terms of displacement, EWS have not been adequately well thought-out, or at least have gotten not as much of interest compared to hazard detection.
- Currently, climate services are under development through the Global Framework for Climate Services (GFCS). In this situation, the lack of effective climate services around the world restricts the current abilities and prospects in the use of climate information for EWS.

7 Conclusions

Early warning in the United Arab Emirates is weak and needs to be consolidated and placed in a framework and therefore benefit from the experiences of countries that have been mentioned in section 3. The authors suggest that there will be a comprehensive development in education and an increase in community awareness of natural disasters during exercises and scenarios common with government institutions and non-governmental organizations. This will strengthen institutional cooperation in the UAE and the role of the media, by disseminating modern methods of warning, such as the internet, television and in various languages in order to benefit the largest possible audience.

The purpose of early warning is to give the opportunity of time to prepare and reduce the confrontation of the victims, as well as the economic and social losses. NCMS in the UAE is mainly responsible for the establishment of an early warning system in the state, in coordination with the concerned authorities, including the Ministry of Interior, in charge of emergency management; public safety and civil defense. This coordination contributes to overcoming the challenges facing early warning in the UAE.

References

[1] Raimondi, M., Natural disasters in developing countries – tropical storm Agatha. KF11, Honduras, June 1, 2010. http://fellowsblog.kiva.org/fellowsblog/2010/06/01/natural-disasters-in-developing-countries-tropical-storm-agatha.

[2] Bogardi, J. & Villagran de León, J.C., Early warning systems in the context of disaster risk management. United Nations University – Institute for Environment and Human Security (UNU-EHS) Görresstraße 1553113 Bonn, p. 23, 2006.

[3] De Marchi, B., Scolobig, A., Delli Zotti, G. & del Zotto, M., Risk construction and social vulnerability in an Italian Alpine region, Task 11: Risk perception, community behaviour and social resilience, Case Study Italy, Floodsite, Directive 2007/60/EC of the European Parliament, 2007.

[4] ISDR, Reducing disaster risks through science: Issues and actions, Report of the ISDR Scientific and Technical Committee, Global Platform for Disaster Risk Reduction, Geneva, 2009. http://www.preventionweb.net/globalplatform/2009/background/documents/SD-03-ENGScientific-Technical-Committee.pdf

[5] Londoño, C.G., Mountain risk management: Integrated people-centred early warning system as a risk reduction strategy, Northern Italy, PhD Thesis, Università degli Studi di Milano-Bicocca, 2011.

[6] UNISDR, Hyogo Framework for Action 2005-2015: Building the Resilience of Nations and Communities to Disasters, 2007. http://www.unisdr.org/we/inform/publications/1037

[7] UNISDR, Early warning – from concept to action: the conclusions of the Third International Conference on Early Warning, EWC III, Bonn, Germany, 32 p., 2006.

[8] Birkmann, J., Seng, D.C. & Suarez, D.C., *Adaptive Disaster Risk Reduction in the Light of Climate Change*, DKKV Publication: Bonn, 2011.

[9] Renn, O. & Walker, K., Lessons learned: a re-assessment of the IRGC framework on risk governance. *The IRGC Risk Governance Framework: Concepts and Practice*, eds. O. Renn & K. Walker, Springer: Heidelberg and New York, pp. 331-367, 2007.

[10] Seng, D.C. & Jury, M., Tropical cyclones in the SW Indian Ocean. Part 2: structure and impacts at the event scale. *Meteorology and Atmospheric Physics*, **106(3)**, pp. 163-178, 2010.

[11] Grasso, V.F., Beck, J. & Manfredi, G., Automated decision procedure for earthquake early warning. *Engineering Structures*, **29(12)**, pp. 3455-3463, 2007.

[12] Basher, R., Global early warning systems for natural hazards: Systematic and people-centered. *Philosophical Transactions of the Royal Society*, **364**, pp. 2167-2182, 2006.

[13] Al-Tamimi, M., Early warning systems and their role in enhancing the ability to cope with crises and disasters in the UAE, Master thesis, Dubai Police Academy, p. 22, 2013.

[14] United Nations, Global Survey of Early Warning Systems. Final Version, http://www.preventionweb.net/files/3612_GlobalSurveyofEarlyWarningSystems.pdf, 2006.

[15] Kapucu, N. & Van Wart, M., The emerging role of the public sector in managing extreme events: lessons learned. *Administration & Society*, **38(3)**, pp. 279-308, 2006.

[16] Caruson, K. & MacManus, S.A., Funding homeland security and hurricane preparedness: Florida's local governments face formidable fiscal pressures. *Municipal Finance Journal*, **27(2)**, pp. 75-93, 2006.

[17] Jin, D. & Lin, J., Managing tsunamis through early warning systems: a multidisciplinary approach. *Ocean & Coastal Management*, **54(2)**, pp. 189-199, 2011.

[18] Collins, M.L. & Kapucu, N., Early warning systems and disaster preparedness and response in local government. *Disaster Prevention and Management*, **17(5)**, pp. 587-600, 2008.

[19] Steinhorst, G. & Vogelgesang, R., The warning management system of the Deutscher Wetterdienst. *Institutional Partnerships in Multi-Hazard Early Warning Systems: A Compilation of Seven National Good Practices and Guiding Principles*, ed. M. Golnaraghi, Springer: London, ch. 5, 2012.

[20] Keeney, H.J., Buan, S. & Diamong, L., Multi-hazard early warning system of the United States National Weather Service. *Institutional Partnerships in Multi-Hazard Early Warning Systems: A Compilation of Seven National Good Practices and Guiding Principles*, ed. M. Golnaraghi, Springer: London, Ch. 6, 2012.

[21] Hasegawa, N., Harada, S., Tanaka, S., Ogawa, S., Goto, A., Sasagawa, Y. & Washitake, N., Multi-hazard early warning system in Japan. *Institutional Partnerships in Multi-Hazard Early Warning Systems: A Compilation of Seven National Good Practices and Guiding Principles*, ed. M. Golnaraghi, Springer: London, Ch. 8., 2012.

[22] Al Ghasyah Dhanhani, H., Evaluation of the Response Capability of the United Arab Emirates (UAE) to the Impact of Natural Hazards, PhD Thesis, University of Bradford, pp. 24-27, 2010.

[23] Al Ameri, F., Implementing an Effective Public Warning System in Abu Dhabi UAE, MSc Thesis, Coventry University, pp. 29-45, 2012.

[24] National Center of Meteorology and Seismology (NCMS), Ministry of Presidential Affairs, UAE, 2013.

Analysis of annual flood peak records in Mexico

C. Escalante-Sandoval & E. Garcia-Espinoza
National Autonomous University of Mexico, Mexico

Abstract

Mexico suffers from the occurrence of floods and droughts. Between 1980 and 2013, floods caused 2,523 fatalities and more than 26 billion dollars of economic damage. The probabilistic assessment of flood risk is made in the form of a flood frequency analysis, which requires the assumption of the stationarity of annual flood peak records. Nonparametric tests were selected to perform homogeneity analysis and to examine the presence of monotonically increasing/decreasing patterns in the data. The 44% of the samples analyzed were considered as non-stationary and the most important changes were detected during the period 1971–1980, where the maximum average annual loss of forest and jungles was experienced. The time of occurrence analysis of the 10 largest flood peaks reveals that in north-western Mexico, the highest values are concentrated in the December-January period. In Southern Mexico, the highest values occur in the October-December period, and for the rest of the country the highest values concentrate in the August-September period. When stationarity cannot be assured, the results of traditional flood frequency analysis could cause fatalities and economic losses, so it is recommended that the National Water Commission take into consideration the theory of non-stationary frequency analysis.
Keywords: stationarity series, flooding, homogeneity test, trend test, flood frequency analysis.

1 Introduction

Mexico extends over a surface area of 1,964 million km^2. The 1,471 catchments identified in the country have been organized into 37 hydrological regions, which are in turn grouped into 13 Hydrological-Administrative Regions.

Mexico's rivers and streams constitute a hydrographic network of 633,000 kilometers. Seven rivers are responsible for 65% of the country's surface runoff: Balsas, Coatzacoalcos, Grijalva-Usumacinta, Pánuco, Papaloapan, Santiago, and Tonalá. The surface area of their watersheds represents 22% of the surface of Mexico. The Balsas and Santiago rivers flow into the Pacific Ocean and the other five flow into the Gulf of Mexico.

Due to its great diversity of climatic and geographical conditions, Mexico is exposed on a daily basis to severe hydro-meteorological events, which, even if they do contribute positively to the increase in water storage in dams and lakes, also cause damage to the population, to infrastructure, to services and production systems. Between 1980 and 2007, floods affected more than 8 million people and caused 2017 fatalities and more than 14.7 billion dollar of economic damage. In this period, cyclones Stan and Isidore were those that affected the largest number of people, whereas Emily, Stan and Gilbert caused the greatest economic damage. Moreover, the events occurred between 2008 and 2009 caused 150 fatalities and 2 billion dollar of economic damage.

The hurricanes that occurred in 2010, such as Alex, Karl and Mathew, mostly affected the states of Campeche, Coahuila, Chiapas, Nuevo León, Oaxaca, Puebla, Tamaulipas and Veracruz. The economic losses that they caused amounted to 7.5 billion dollar, and the death toll was 199.

In September 2013, at least 157 people died in Mexico by the simultaneous occurrence of two powerful storms, one on its east coast (hurricane Ingrid) and another on its west coast (tropical storm Manuel). The economic losses amounted to 1.8 billion dollar.

The damage associated with floods are increasingly more serious due to the location of irregular human settlements in areas close to rivers, the lack of enforcement of land use regulations, and deforestation in the upper portions of river basins; the effect of the latter is reflected in an increase in runoff and the transportation of soils and sediments towards the lower portions.

Flood damage reduction consists in adopting structural and non-structural measurements. Structural methods imply designing works such as levees, floodwalls, dams, dredging and channelization. Moreover, non-structural measures can be the relocation, flood proofing, buy-out or acquisition, and flood preparedness. In particular, the structural measurements require risk analysis of floods as design criteria. The probabilistic assessment of flood risk is made in the form of a flood frequency analysis.

The assumption of stationarity of annual flood peak records is very important in risk analysis for flood hazards. The validity of such assumption has been challenged due to the natural variability and induced anthropogenic changes in the climate system [1].

The annual flood peak time series is stationary if it is free of periodicities and abrupt changes [2]. In flood frequency analysis when stationarity cannot be assured, the estimated event for design purposes could lead to catastrophic consequences in case of an underestimation and economic losses in case of an overestimation. Ouarda and El-Adlouni [3] provide a broad discussion of non-stationary frequency analysis in hydrology.

As already mentioned, flood events are responsible for numerous fatalities and large economic damage, so a detailed analysis of the flood peak records will allow a better understanding of that possible temporal change.

This study focuses on investigating the presence of temporal non stationarities in the flood peak records. To address this problem, annual maximum flood peak records are used collected from 205 gauging stations with a record of at least 50 years over the entire Mexican territory.

The validity of the stationarity assumption is checked in terms of both abrupt changes in the mean and variance of the flood peak distribution and monotonic trends. Some nonparametric tests were used to perform homogeneity analysis for the detection of abrupt changes in the mean and variance of the flood peak distribution, and to examine the presence of monotonically increasing/decreasing patterns in the data. For this purpose, the methodology suggested by Villarini *et al.* [4] was adopted according to which corrections to account for serial correlation or long-term persistence are not considered.

2 Data and methods

2.1 Data

The National Water Commission has available information from 990 gauging stations located all over Mexican territory. However, in this study 205 stations are only considered, with a record of at least 50 years of annual maximum instantaneous peak discharge (as shown in fig. 1).

Figure 1: Location of the 205 gauging stations included in this study.

2.2 Change-point analysis

Four statistical tests were used to test homogeneity of the mean: the standard normal homogeneity test [5], the Buishand range test [6], the Pettitt test [7] and the Von Neumann ratio [8]. Moreover, the F-test [9], Levene's test [10], the Brown–Forsythe test [11] and Bartlett's test [12] are used to verify the assumption of homoscedasticity or homogeneity of variances. Both mean and variance homogeneity tests were assessed as follows:

The series that rejects one or none null hypothesis under the four tests at 5% significance level are considered to indicate homogeneity.

The series that reject two null hypotheses of the four tests at 5% significance level indicate series that have the inhomogeneous signal and should be critically inspected before further analysis.

When there were three or all tests rejecting the null hypothesis at 5% significance level they were considered as indicating non homogeneous conditions.

2.3 Trend analysis

The presence of monotonic trends was examined by means of two nonparametric monotonic trend tests, Mann–Kendall and Spearman rho test [13]. For a given record, the tests were performed on the entire record if the series was considered homogeneous. If a change-point in mean was detected, the analysis was performed on the two subseries (before and after the change-point).

3 Results and discussion

The change-point analysis in mean and variance was performed using the statistical tests aforementioned. Out of 205 stations, 41% of the stations exhibiting a change point in mean, 45% in variance, and 41% a change-point both in mean and variance. Maps with stations showing a change-point are presented in figs 2 and 3.

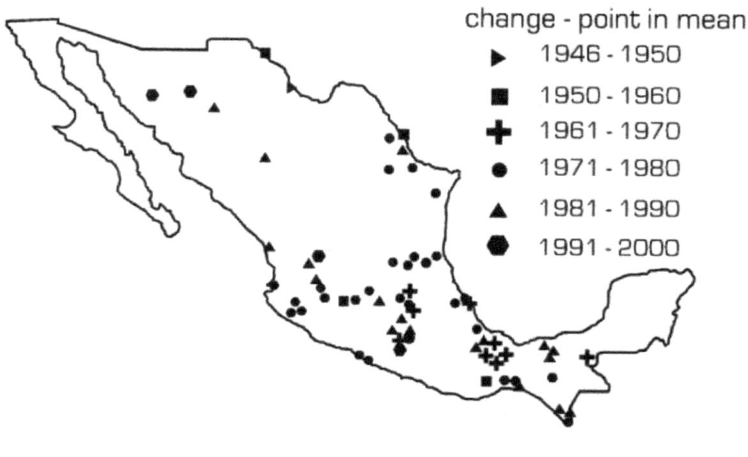

Figure 2: Stations with a change-point in mean.

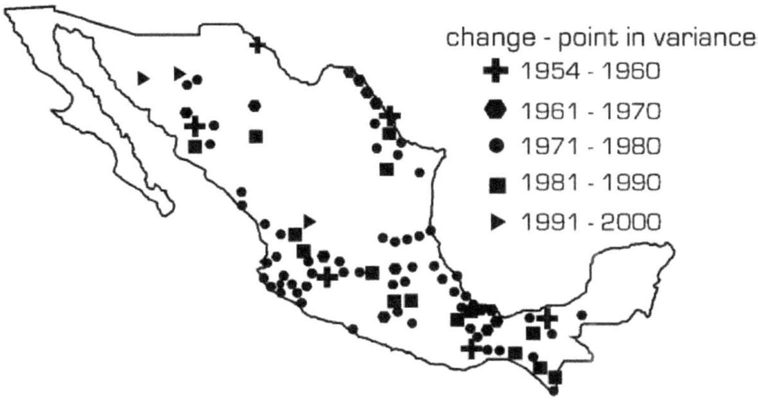

Figure 3: Stations with a change-point in variance.

The most important changes were detected during the period 1971–1980. These change-points can be related to the accelerated process of deforestation experienced in the past decades. The disappearance of forests in Mexico is mainly attributed to the increased area devoted to large-scale agriculture and cattle-raising, particularly in tropical and sub-tropical regions.

The decadal average annual loss of forest and jungles is listed in table 1.

Table 1: Decadal average annual loss of forest and jungles in Mexico.

Decade	Loss (ha)
1951–1960	400,000
1961–1970	600,000
1971–1980	700,000
1981–1990	400,000
1991–2000	300,000
2001–2010	200,000

The maximum average annual rate was experienced in the period 1971–1980. In later years, Mexico has been reducing the deforestation rate applying its plans to conserve and restore forests and jungles.

After this analysis, 121 stations could be considered homogeneous in mean and variance (as shown in fig. 4). Most of the inhomogeneous stations are located in three, well defined regions. The first of them is located in the central portion of Mexico and is called “The Lerma Chapala basin”. This is a strategic region for the country development. Nevertheless, the dynamics of its socioeconomic growth during the last decades has triggered severe water resources problems and conflicts as well as environmental degradation and deterioration. The basin comprises portions of 5 Mexican states: Jalisco, Guanajuato, Michoacan, Queretaro and Mexico. The second one is located in northern Mexico and is called “The Rio

Bravo basin"; it covers three US states and four Mexican states: Chihuahua, Coahuila, Nuevo Leon and Tamaulipas. In this basin, the rapid population growth and consequent economic development and land-use changes are pushing the limits of environmental sustainability and quality. During the last century, a number of large reservoirs were built on the Río Bravo and its major tributaries and extensive well fields were drilled in the basin's aquifers. The dams have greatly reduced the downstream flow of the main stem and its tributaries. The last region is located in southern Mexico, and covers portions of three Mexican States: Veracruz, Tabasco and Chiapas. This region has the biggest regional quantity of renewable water resources, a relatively low population and contribution to the GDP. The severe land-use changes in this region are attributed to the increased area devoted to large-scale agriculture and cattle-raising.

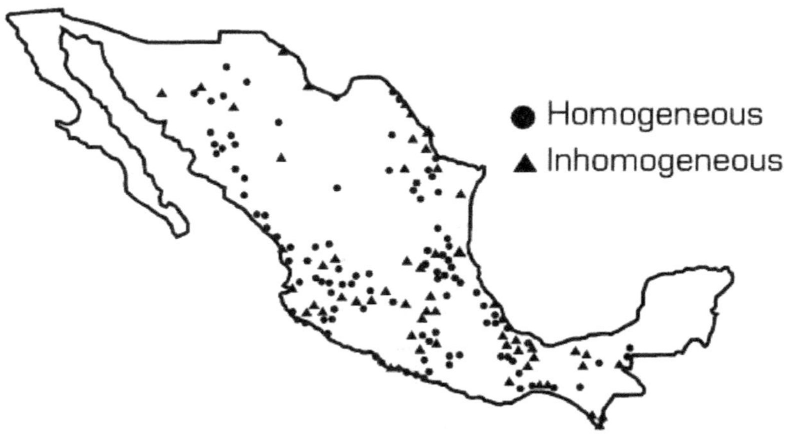

Figure 4: Map showing the final results of the homogeneity tests.

Trend analysis is performed on the entire time series if there was not statistically significant change point in mean (there were 121 stations fulfilling this requirement). From this set of stations, 4 show a negative trend and only 2 a positive trend (as shown in fig. 5). Moreover, if a change-point in mean was detected, analysis was performed on the two subseries, before and after the change-point. In this case (84 stations), 10 stations showed positive trend before change-point, as shown in fig. 6, and 15 indicate negative trend after change-point, as shown in fig. 7. In most cases, the change-points rather than monotonic trends were responsible for most violations of the stationarity assumption.

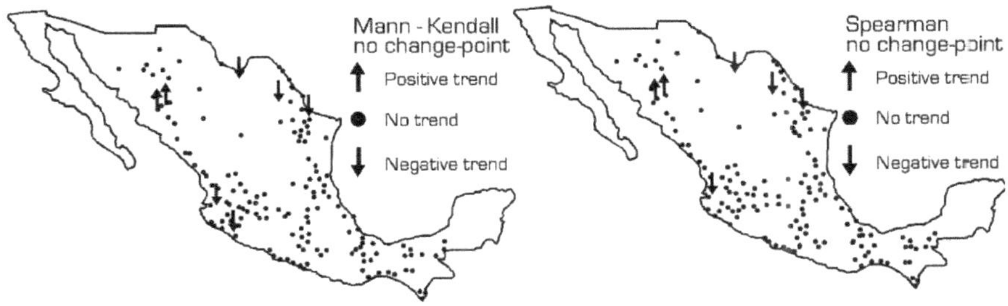

Figure 5: Results of the Mann–Kendall test and Spearman test for the series without change-point in mean.

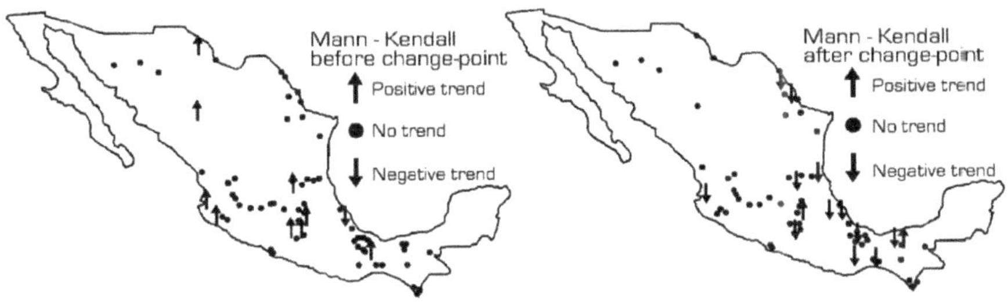

Figure 6: Results of the Mann–Kendall test before and after the change-point in mean.

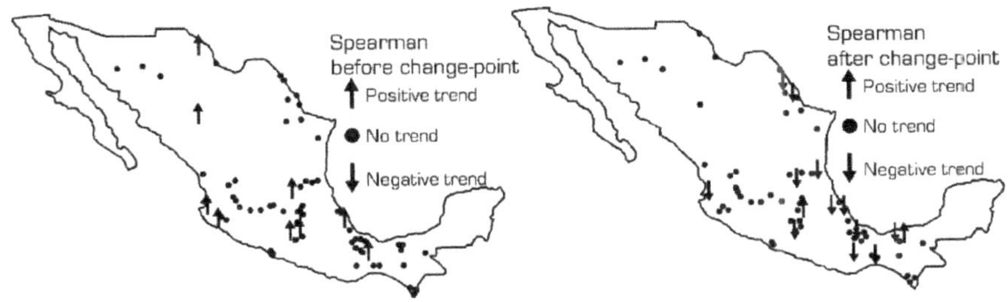

Figure 7: Results of the Spearman test before and after the change-point in mean.

The probabilistic assessment of flood risk was made in the form of a flood frequency analysis. This analysis assumes that data are independent and identically distributed, which implies that it must meet the statistical criteria of independence, stationarity, and homogeneity (there were 115 stations fulfilling this requirement). For this purpose, unimodal and mixed distributions were used. The selection of the probability distribution that best fits the sample data is based on a goodness-of-fit criterion [17]. The best fits are distributed according to the percentages shown in table 2. The spatial distribution of these fits is presented in fig. 8.

Table 2: Percentage of occurrence of the best distribution of fit.

Distribution	%
Pearson Type III	55
Mixed Gumbel	21
Generalized Extreme Value	11
Gumbel	4
Gamma 2 parameters	3
Lognormal 3 parameters	3
Lognormal 2 parameters	3

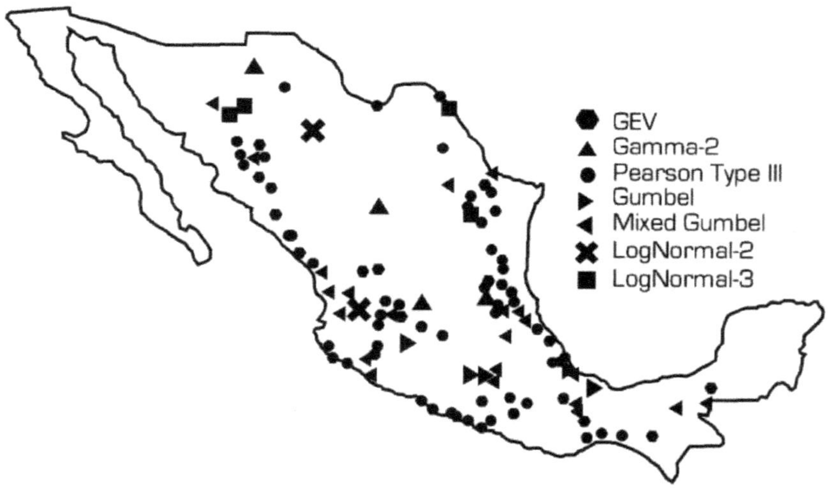

Figure 8: Location of each gauging station, each associated with its best probability distribution of fit.

Now, it is possible to estimate the quantiles associated with different return periods, which are required in flood risk management. However, it was also necessary to know the time of occurrence of the flood peak. To accomplish this objective, the percentage of occurrence of the 10 largest peaks from January to December was plotted as shown in figs. 9 and 10. In Northwestern Mexico, the highest values are concentrated in the December-January period. In Southern Mexico, the highest values occur in the October-December period, and for the rest of the country the highest values concentrate in the August-September period.

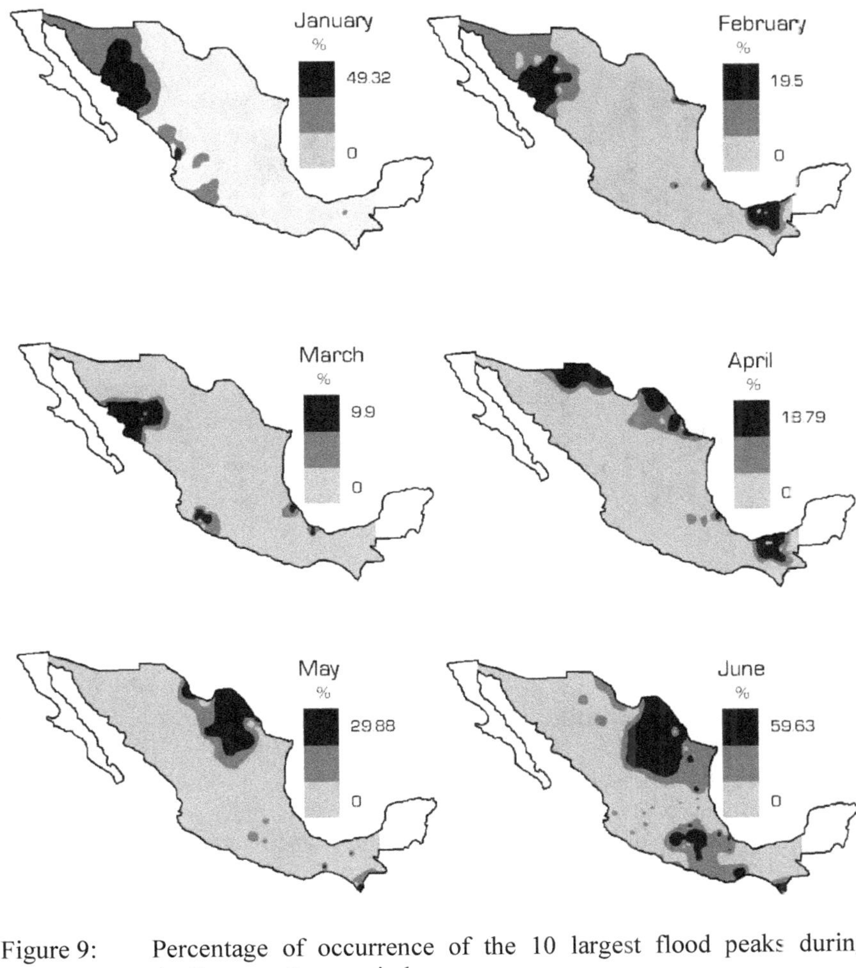

Figure 9: Percentage of occurrence of the 10 largest flood peaks during the January–June period.

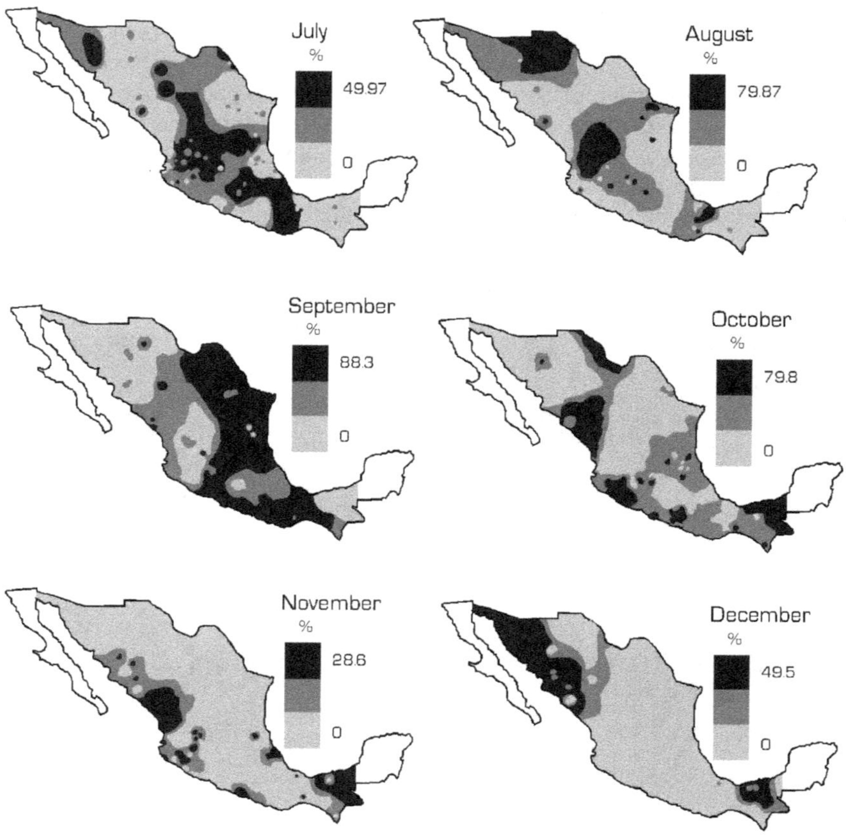

Figure 10: Percentage of occurrence of the 10 largest flood peaks during the July–December period.

4 Concluding remarks

Mexico is periodically subject to meteorological vents that cause serious damage in large portions of its territory and it is expected that they will be even more frequent and intense due to the effects associated with climate change.

As the drainage basin scale, consideration of flood risk plays an important role in planning water infrastructure projects. The probabilistic assessment of flood risk is made in the form of a flood frequency analysis, which requires the assumption of stationarity of annual flood peak records.

Four nonparametric tests were used to perform homogeneity analysis for the detection of abrupt changes in the mean and another four for the variance. Moreover, two additional nonparametric tests were applied to examine the presence of monotonically increasing/decreasing patterns in the data. Out of 205 stations, 41% of the stations exhibit a change point in mean, 45% in variance, and 41% a change-point both in mean and variance.

The 44% of the analyzed samples were considered non stationary and the most important changes are detected during the period 1971-1980. In the vast majority of the cases, violations of the stationarity assumption are associated with change-points rather than monotonic trends. These non-stationarities could be associated with anthropogenic effects, such as the population growth, unordered urban settlements, the deforestation and degradation of catchments.

From the total of 205 stations, 115 did not present statistically significant change-points and trend. The trend distribution is described in terms of its cumulative distribution function. The 55% of the samples were best fitted by the Pearson Type III distribution, followed by the Mixed Gumbel distribution (21%) and the Generalized Extreme Value Distribution (11%).

The time of occurrence analysis of the 10 largest flood peaks reveals that in north-western Mexico, the highest values are concentrated in the December–January period. In Southern Mexico, the highest values occur in the October–December period, and for the rest of the country the highest values concentrate in the August–September period.

When stationarity cannot be assured, the results of traditional flood frequency analysis produce representations that might cause fatalities and economic losses, so it is recommended that Mexican authorities take into consideration the non-stationary frequency analysis theory.

With this kind of analysis, authorities can be able to substantially improve the emergency plans for rivers that may cause damage by overflowing. As a complement to this, the reforestation program in the upper parts of river basins and eroded areas must be strengthened. This will contribute to reducing possible floods in the lower parts of river basins.

References

[1] Milly, P., Betancourt, J., Falkenmark, M., Hirsh, R., Kundzewicz, Z., Lettenmaier, D. and Stouffer, R., Stationarity is dead: whiter water management? *Science*, **319**, pp. 573-574, 2008.

[2] Matalas, N., Stochastic hydrology in the context of climate change. *Climate Change,* **37**, pp. 89-101, 1997.

[3] Ouarda, T. & El-Adlouni, S., Bayesian non-stationary frequency analysis of hydrological variables. *Journal of the American Water Resources Association*, **47(3)**, pp. 496-505, 2011.

[4] Villarini, G., Smith, J., Baeck, M. & Krajewski, W., Examining flood frequency distributions in the Midwest US. *Journal of the American Water Resources Association*, **47(3)**, pp. 447-463, 2011.

[5] Alexandersson, H., A homogeneity test applied to precipitation data. *Journal of Climate*, **6**, pp. 661-675, 1986.

[6] Buishand, T., Some methods for testing the homogeneity of rainfall record. *Journal of Hydrology*, **58**, pp. 11-27, 1982.

[7] Pettitt, A., A non-parametric approach to the change-point problem. *Journal of Royal Statistical Association*, Series C, **28(2)**, pp. 126-135, 1979.

[8] Neumann, J. von, Distribution of the ratio of the mean square successive difference to the variance. *Annals of Mathematical Statistics*, **13**, pp. 367-395, 1941.

[9] Rees, D.G., *Essential Statistics*, 4th ed., Chapman & Hall/CRC, 2001.

[10] Levene, H., *Contributions to Probability and Statistics: Essays in Honor of Harold Hotelling*, Stanford University Press, 1960.

[11] Brown, M. & Forsythe, A., Robust tests for equality of variances. *Journal of the American Statistical Association,* **69**, pp. 364-367, 1974.

[12] Snedecor, G. & Cochran, W., *Statistical Methods,* 8th ed., Iowa State University Press, 1989.

[13] Kundzewics, Z. & Robson, A., Change detection in hydrological records – A review of the methodology. *Hydrological Sciences Journal*, **49(1)**, pp. 7-19, 2004.

[14] Douglas, E., Vogel, R. & Kroll, C., Trends in floods and low flows in the United States: Impact of serial correlation. *Journal of Hydrology*, **240**, pp. 90-105, 2000.

[15] Yue, S., Pilon, P. & Cavadias, G., Power of the Mann-Kendall and Spearman's rho test for detecting monotonic trends in hydrological series. *Journal of Hydrology*, **259**, pp. 254-271, 2002.

[16] Yue, S. & Wang, C.Y., Applicability of prewhitening to eliminate the influence of serial correlation on the Mann-Kendall test. *Water Resources Research*, **38(6)**, 4-1–4-7, 2002.

[17] Kite, G., *Frequency and Risk Analyses in Hydrology*, Water Resources Publications: Fort Collins, 1988.

A preliminary flood-risk assessment of municipalities located in headwater basins of Slovakia based on an integrated approach

Ľ. Solín, P. Skubinčan & M. Madajová
Institute of Geography, Slovak Academy of Sciences, Slovakia

Abstract

The aim of preliminary flood-risk assessment, in general, is to identify the spatial variability of flood risk; this creates a rational basis for its detailed assessment and management. The traditional engineering approach to the evaluation of flood risk primarily aims at assessing the ability reached by a river to transport the discharge of the specified probability of non-exceedance. In this paper, an integrated approach addresses this issue. The flood-risk assessment of the municipalities situated in the headwater basins of Slovakia is based on a combination of flood hazard and expected negative consequences. The flood hazard in headwater basins is understood as a potential – that is, the disposition of the basin to suffer from flood – which is influenced by, besides rainfall, the attributes of the basin. The negative consequences are influenced by the vulnerability and resilience of the municipalities. Vulnerability is characterized by six proxy variables: the materials the houses are built from, the period in which they were built, the age structure of the population, ethnicity, completeness of the family and category of house. The ability of the municipalities to recover after a flood is expressed by the flood insurance rate. The municipalities are classified into five categories of flood risk based on the combination of variables reflecting the flood hazard, vulnerability and resilience.
Keywords: integrated flood-risk management, flood risk, flood hazard, vulnerability, flood insurance, multi-criteria analysis.

1 Introduction

Currently, a change in paradigm of flood protection is witnessed in many countries (e.g. [1–3]). From the exclusive application of engineering flood management (mostly by means of reducing the extent of flooding), flood protection gradually shifts towards integrated flood-risk management [4, 5], which, in addition to reducing the scope of flooding, also focuses on the reduction of susceptibility to damage and mitigation of the impacts of flooding. The new paradigm requires a change of approach to the evaluation of flood risk.

The traditional engineering approach to flood-risk assessment constitutes specification of the magnitude of the discharges Q_{max} of the non-exceeding probability F. A preliminary evaluation primarily aims to assess the state of the river sections in municipalities (the shape, the size of the trough, and the stability of the banks) with regard to their ability to transport the flow rate of the specified probability of non-exceedance.

However, integrated flood-risk management addresses the assessment of flood risk in a more complex way. In addition to flood hazard, significant attention is paid to the assessment of the vulnerability of society to the floods. This change in the assessment of flood risk in particular reflects a change in understanding the concept of risk. Unlike the engineering approach that understands the concept of risk in relation to the threat of flooding only (flood risk = flood hazard), integrated flood-risk management understands the risk as the expected losses (of lives, persons injured, property damaged and economic activity disrupted) due to a particular hazard for a given area and reference period [6] and, therefore, flood risk = (flood hazard)×(negative consequences). Flood hazard can be expressed in the form of probability (magnitude of a flood in terms of its non-exceedance probability) or in the form of potential (the possibility of the basin suffering from floods). The negative effects of a flood depend on the vulnerability of objects of the social, economic and environmental systems to floods in a given area (e.g. [7–9]).

This paper deals with the preliminary integrated flood-risk assessment of municipalities located in the headwater basins of Slovakia. The aim of the preliminary flood-risk assessment is to identify the spatial variability of flood risk and thus to create a rational basis for its detailed assessment and management. Emphasis is laid on municipalities in small basins because Slovakia is a predominantly mountainous country (mountains account for nearly 60% of the total area of Slovakia). Almost 47% of all 2,928 municipalities are located in headwater basins in mountainous areas. In contrast, only 24% of municipalities are located in close vicinity to medium and large streams flowing in lowland areas. Nevertheless, an assessment of flood-risk management in small mountain catchments has not received sufficient attention. Flood management in Slovakia primarily aims to protect areas located along the middle and lower sections of the rivers in the lowlands with high concentrations of the population and economic activities. However, this preference for flood protection for residential areas located mainly in the lowland areas is not fully justified. In the period 1996–2006, 47% of all municipalities in Slovakia were affected by flooding. Of these flood events, 67% occurred in the municipalities localized in small headwater basins [10]. In the small mountain river basins, flood hazard, in addition to watercourse flooding, is often also caused by the concentration of surface runoff in the slope areas during short but intense rainfall. Water which is on the surface of the soil gradually concentrates into ever larger streams and eventually gives rise to a large stream of water that rolls downhill and damages objects that are in its way.

Floods in small basins are therefore a significant phenomenon in Slovakia and corresponding attention must be paid to flood-risk management in these areas. Aware of this fact, work systematically started on the flood-risk assessment process, based on an integrated approach in municipalities located in headwater basins. First, a database of the attributes of small river basins was created in GIS [11]. Second, the occurrence of flood events was added to the database [10]. Later, the effects of the basin attributes and land cover changes in the period 1990–2006 on the occurrence of flood events were analysed [12, 13]). The social vulnerability of the municipality populations with respect to floods was also evaluated [14], as well as their ability to cope with the negative consequences of floods [15].

The aim of the paper is to move to the integrated assessment of flood risks of municipalities located in small basins by the combination of results achieved in relation to the assessment of the individual components of flood risk.

2 Data

Preliminary flood-risk assessments of the municipalities located in headwater basins were made using the following data: a) a digital layer of boundaries of the headwater basins, b) a database of basin attributes, c) a digital layer of municipality centroids and boundaries and d) a database of the socio-economic attributes of the municipalities. The digital layer of the headwater basins, which functions as the basic spatial unit, was created by Solín and Grešková [16], who modified the borders of the river basins which were digitized from the 1:50,000 scale water management maps by the Slovak Environmental Agency (SEA) in the period 1997–1998. The digital layer contains almost 4,587 small basins, with the area of a single basin varying from 0.04 to 150 km^2. For the purposes of this study, a basic set of headwater basins was selected, which included 1,678 basins, mostly in mountainous areas of Slovakia, with a basin area in the range of 5–150 km^2. A database of their morphometric, geological and hydrological attributes, as well as soil texture parameters and characteristics of landscape cover was created, using the digital elevation model (CORINE Land Cover [17]), various digitized thematic maps and the outcomes of the processing of hydrological and rainfall data [11]. A digital layer of the centroids and boundaries of the municipalities was provided by the SEA. Data used in the process of assessing the vulnerability of the municipalities' economic and social systems were primarily the 2001 census data. The ability to cope with the negative consequences of floods and the capabilities of communities to restore their original state was expressed through the level of insurance against the risk of flooding. The data about flood insurance policies, along with data about flood insurance events was provided by the Allianz-Slovak Insurance Company.

3 Flood-risk assessment methodology

In general the conceptual and methodological framework of flood-risk assessment is based on a combination of the flood hazard and the vulnerability of the social, economic and environmental systems of a particular area. There are two groups of methods for determining the level of flood risk. The first group consists of methods expressing flood risk in an absolute way by *average annual damage value* (cf. [18]). Expressing flood risk in an absolute way combines the expected losses from all levels of hazard probability. The second group includes the method expressing flood risk in a relative way by ordinal scale, that is, dimensionless values of the flood hazard and the vulnerability of the economic, social and environmental systems are aggregated and then ranked into classes expressing high, moderate or low levels of risk. This process is the core of spatial multi-criteria decision analysis (cf. [19, 20]). There are two types of multi-criteria decision analysis (MCDA): multi-attribute decision-making (MADM) and multi-objective decision-making (MODM).

In this paper, due to the fact that the research is carried out at the national level and the subject of an assessment is a set of 1,878 municipalities, the ordinal scale approach based on MADM was applied to assess flood risk. The following model represents this approach:

$$U_i = \sum_j w_j u_{ij} \tag{1}$$

where U_i is the flood risk score of municipality i, u_{ij} is the standardized score of the variable j corresponding to the flood hazard and vulnerability of the economic and social systems of municipality i and w_j is a weighting that represents the effect of variable j on the flood risk.

The variables were standardized using the maximum score approach [20]. The transformation equation is:

$$x'_{ij} = \frac{x_{ij}}{x_j^{\max}} \tag{2}$$

where x'_{ij} is the standardized score, x_{ij} the criterion score and $x_j^{\max}$ the maximum criterion score.

As far as the weighting of variables is concerned, the assumption that all variables have the same weight was adopted. Using this method, the value of the flood-hazard index and the value of the vulnerability index were determined. Their subsequent aggregation later formed the flood-risk index.

Flood hazard can be expressed either on a probability or potential base. In the first case, the process of flood-hazard assessment constitutes specification of the magnitude of discharges $Q_{\max}$ of the non-exceeding probability F and the mapping of the corresponding flooding. In the second case, the flood hazard is expressed by means of a potential – that is, the disposition of the basin to suffer from flood – which is influenced by the basin attributes [12, 21]. In this paper, the flood-hazard assessment was made on the basis of the basin potential. The flood hazard of the municipality is then derived from the flood-hazard potential of the basin in which the municipality is located. When deciding on the approach to the evaluation of the flood hazard, account was taken of both the objective of the preliminary flood-risk assessment (the identification of the spatial variability of the flood risk) as well as the scale of the assessments (national scale), which led to choosing the potential-based form.

The flood vulnerability assessment expresses propensity to damage, resistance to flood and capacity of post flood recovery only in terms of the properties of the social, economic and environmental systems of the territory [22–25]. In this context, vulnerability means a potential. It does not contain the element of flood-hazard exposure; hence it is *hazard-independent* [26, 27]. Vulnerability indices are established through a combination of normalized variables that characterize the vulnerability of the economic and social systems of municipalities.

4 Results

4.1 Assessment of the flood hazard

The evaluation of a basin's potential for the occurrence of a flood is based on the analysis of hydrographs. Figure 1 shows the hydrographs of the average daily flow of two river basins, which are different in terms of their attributes. The hydrograph in fig. 1(a) is from the river basin geologically dominated by flysch rocks, that is, the dominant drainage process in this basin is direct runoff (BFI – base flow index = 0.26) which usually creates sudden and steep flow waves. In contrast, fig. 1(b) represents a basin dominated by crystalline slate, and suggests that the predominant drainage process in this river basin is basic runoff (BFI = 0.68). For this form of runoff, typical flow waves are shallow and long lasting.

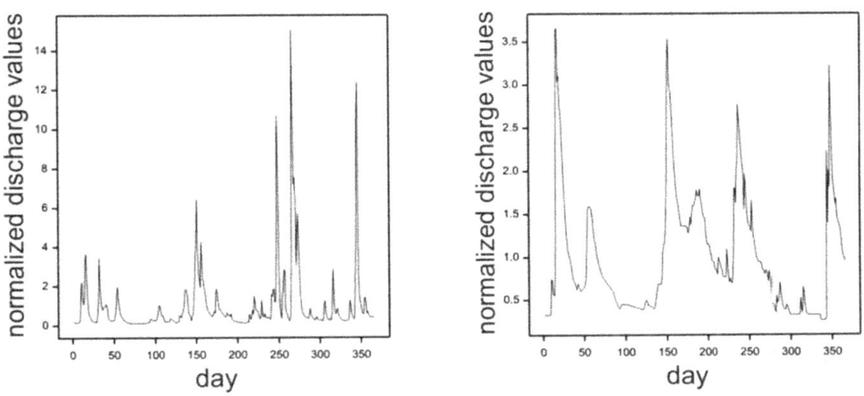

(a) (b)

Figure 1: Examples of different hydrological responses: (a) River Ladomírka, (b) River Rimava.

From this different hydrological response, it can be concluded that the flood threat in the first basin is much higher than that in the second. The results of the work of Solín [12] have shown that the permeability of the soil texture has the most significant impact on the hydrological response of all basins' attributes. The permeability of the soil texture of the river basin was established on the individual percentage of each soil texture class within the basin area. The second attribute that affects the hydrological response is deforestation of the basins.

While testing the different combinations of both attributes, it was proven that the optimal grouping of headwater basins is into three regional flood-hazard classes. Each class was assigned ordinal values of flood hazard: 3 as high, 2 as medium, and 1 as low. Standardized ordinal values reflect the flood-hazard index. By overlaying the layer of flood-hazard classes of the basins with the layer of centroids representing individual municipalities, it was possible to establish the level of flood hazard for each municipality (corresponding to the flood-hazard index of the basin in which it was located) (fig. 2).

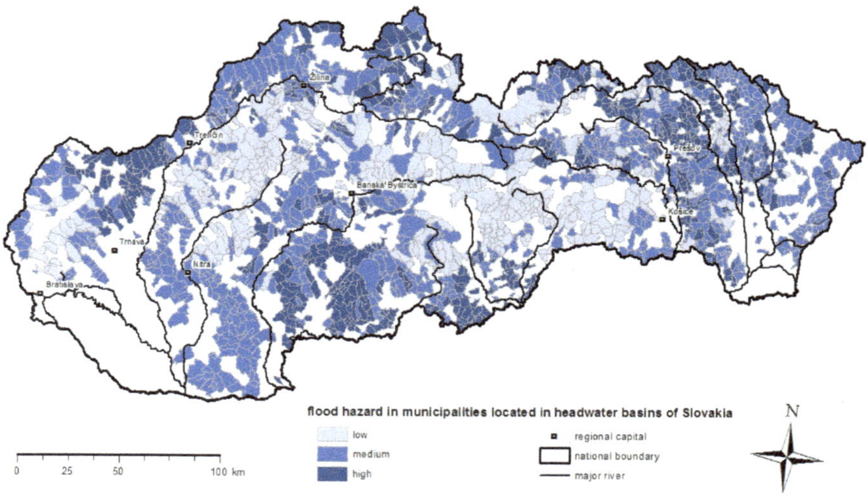

Figure 2: Flood hazard of municipalities located in headwater basins.

The diversity of the hydrological responses as a result of the basin attributes is reflected in the different frequencies of the flood events in the municipalities (fig. 3). A flood event occurs when the mayor of the municipality announces a 3rd degree flood alert.

4.2 Assessment of the social and economic vulnerability

The assessment of the three components of the vulnerability to floods, that is, the susceptibility of the economic system to damage, the resistance of the social system to floods and their capacity for recovery after the flood is based on the results achieved by Solín [14] and Solín *et al.* [15]. In the first paper, the social and economic vulnerability to flooding of the Basic Settlement Units (BSU) is analysed. For the purpose of this study, BSUs were aggregated into higher territorial units – municipalities.

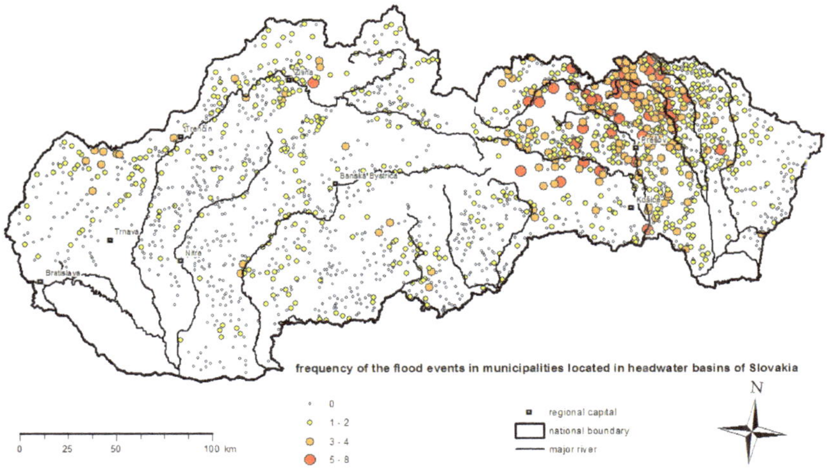

Figure 3: Frequency of the flood events in municipalities in the period 1996–2010.

Data on the variables and attributes of the economic and social systems used for vulnerability analysis are summarized in table 1.

Table 1: Proxy variables of economic and social vulnerability.

System	Variable	*Attributes of variable (percentage)*
Economic	Material of bearing walls	*Earthen houses, other and unspecified materials*
	Date of construction	*Houses before 1945*
Social	Age structure	*People over the age of 65*
	Ethnicity	*Roma*
	Family completeness	*Incomplete*
	Category of family house	*Houses of the 3rd and 4th category*

Considering the materials the houses were built from and the year of construction of the family houses forms the *economic system susceptibility index*. A combination of three variables, age structure of the population, ethnicity and completeness of family, forms the *social system resistance index*. The capacity of municipalities to recover damaged property and return to their pre-flood state depends on the financial situation of the people living there. Financial deprivation was assessed indirectly using a categorization of the family houses in which people live. This variable is used to form the *social system resilience index*. By aggregating partial indices together, an *index of total vulnerability* of the municipality to floods is obtained (fig. 4).

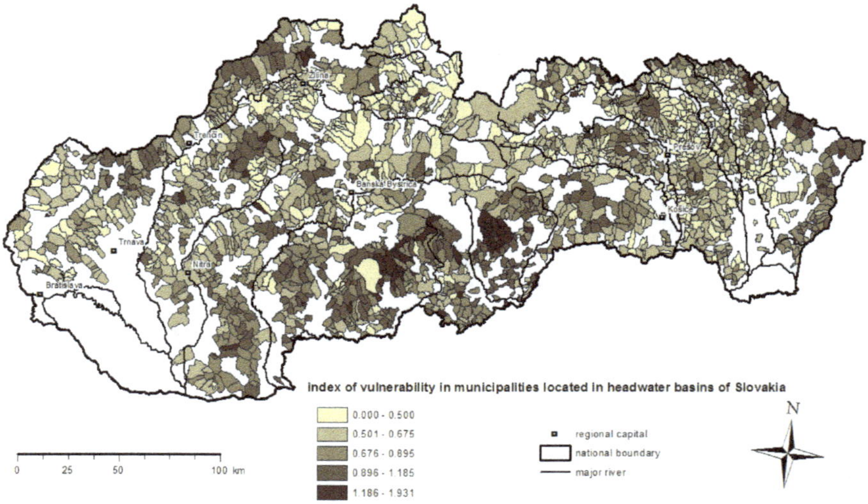

Figure 4: Vulnerability index of municipalities located in headwater basins.

The ability to cope with the negative consequences of the floods also depends on the level of insurance against damage caused by flooding. Empirical data on the flood-risk insurance of individual municipalities was analysed by Solín *et al.* [15]. In the past decade, a dramatic decrease in the population's interest in insuring their property against flooding has occurred. Only 58% of policies active in 2002 were still in force in 2011. The mean level of properties insured against flooding in municipalities situated in headwater basins decreased from 60% in 2002 to 30% in 2011. Apparently, these values have to be interpreted with some caution, as in this case, data from only one, although by far the largest, insurance company in the market were considered. Accounting for the market share of other insurance companies in the field of non-life insurance, it can be presumed that the current overall level of flood-risk insurance in Slovakia is around 50%. The insurance rate is expressed as the ratio of the number representing flood-risk policies in individual municipalities to the number of residential buildings (single-family houses and flats). Standardized values of the insurance rate form the *index of flood impact mitigation* (fig. 5).

4.3 Flood-risk assessment

The level of overall flood risk is expressed by the index, which is the aggregate of the values of the indices reflecting flood hazard, vulnerability and level of insurance. Municipalities located in headwater basins were, according to the flood-risk index, classified into five ranks: very high, high, medium, low and very low levels of flood risk (fig. 6).

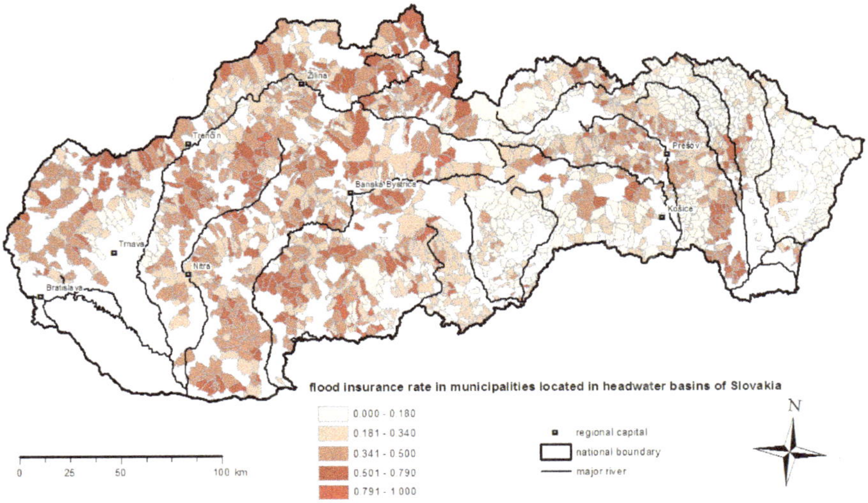

Figure 5: Flood insurance rate in municipalities located in headwater basins.

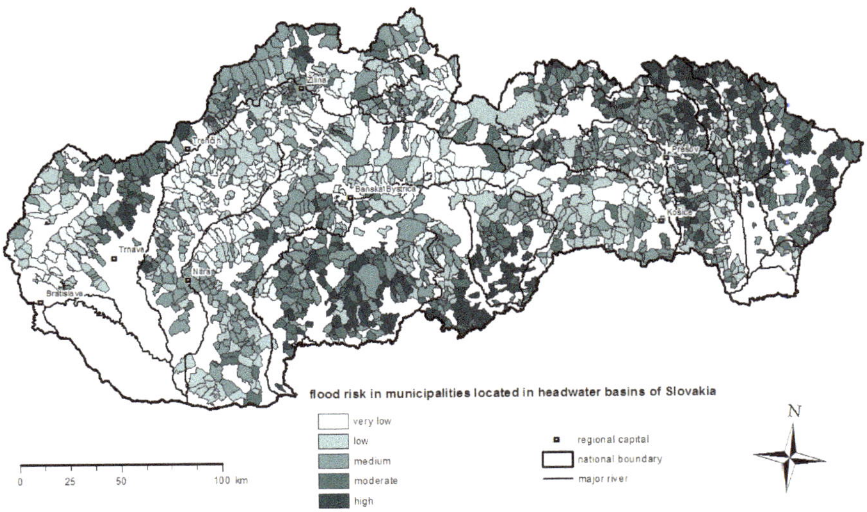

Figure 6: The level of flood risk of municipalities located in headwater basins.

As already mentioned above, integrated flood-risk management is a combination of three primary strategies, which aim to reduce the extent of flooding, thus reducing vulnerability and mitigating the negative consequences of the floods.

Flood-risk assessment based on an integrated approach is therefore a rational basis for the application of a spatially differentiated approach in designing the optimal strategy for flood-risk management plans. Share of the primary strategies within the framework of the optimal "general" strategy will depend on the level of flood risk. With an increasing level of flood risk, flood-risk management will move from the emphasis on the strategy eliminating the negative consequences of flood towards strategies reducing vulnerability and the extent of flooding. The specification of the optimal strategy for each class of flood risk is a significant step forward in the management of flood risks in comparison to the uniform application of the strategy reducing only the scope of flooding used in the traditional approach.

The result of the preliminary integrated flood-risk assessment is an effective way to select areas for subsequent creation of probabilistic flood hazard maps and maps of flood risk, which are the basis for the assessment of the financial effectiveness of the proposed flood protection measures.

The expression of the results of the preliminary integrated flood-risk assessment, using the ordinal scale rather than the binary (yes/no) classification of the engineering approach, creates room for making flood-risk management exhaustive, that is, for each municipality an optimal flood management strategy can be determined based on its flood-risk level. In the case of a binary array of flood-risk assessment, the aim of flood-risk management concentrates only on municipalities with segments of river that have potentially significant flood risk, while neglecting other nearby municipalities with no potentially significant risk.

5 Conclusions

A shift in the flood protection paradigm from engineering flood management to integrated flood-risk management also requires a new approach to evaluate flood risk. Based on the integrated approach, in this article, a preliminary flood-risk assessment of the municipalities located in headwater basins has been presented. Based on the mutual combination of the attributes of flood hazard, social and economic vulnerability and the ability to cope with the negative consequences of floods, municipalities were classified into five categories of flood risk. In comparison with the engineering approach, the results of the preliminary integrated flood-risk assessment create a much more efficient basis for the rational management of flood risk.

Acknowledgements

This paper is one of the outputs of the 2/0091 project *Flood Risk in Communities of Slovakia,* financially supported by the Scientific Grant Agency of the Ministry of Education, Science, Research and Sport of the Slovak Republic and the Slovak Academy of Sciences (VEGA).

References

[1] Brown, J. & Damery, S., Managing flood risk in the UK: towards an integration of social and technical perspectives. *Transactions/Institute of British Geographers*, **27**, pp. 412–426, 2002.

[2] Fuchs, S., Susceptibility versus resilience to mountain hazards in Austria – paradigms of vulnerability revisited. *Natural Hazards and Earth Systems Sciences*, **9**, pp. 337–352, 2009.

[3] Hassel, H., Risk and vulnerability analysis in practice: evaluation of analyses conducted in Swedish municipalities. *Natural Hazards*, **63**, pp. 605–628.

[4] APFM (The Associated Programme on Flood Management), *Integrated Flood Management*. Concept paper. The Associated Programme on Flood Management, World Meteorological Organization and Global Water Partnership, Geneva, 2004.

[5] Ale, B., Aven, T. & Jogejan, R., Review and discussion of basic concepts and principles in integrated risk management. *Reliability, Risk and Safety: Theory and Applications,* eds R. Briš, C. Guedes Soares & S. Martorell, Taylor & Francis Group: London, pp. 421–427, 2010.

[6] UN (United Nations), *Internationally Agreed Glossary of Basic Terms Related to Disaster Management Risk*. UN Department of Humanitarian Affairs, Geneva, 1992.

[7] Wisner, B., Blaikie, P., Cannon, T. & Davis, I., *At Risk: Natural Hazards, People's Vulnerability, and Disasters*. 2nd edition. Routledge: London, 2004.

[8] Messner, F. & Meyer, V., *Flood Damage, Vulnerability and Risk Perception – Challenge for Flood Damage Research*. Umweltforschungszentrum Lepizig-Halle, Discussion papers 13, 2005.

[9] Fuchs, S., Bierkman, J. & Glade, T., Vulnerability assessment in natural hazard and risk analysis: current approaches and future challenges. *Natural Hazards* **64**, pp. 1969–1975, 2012.

[10] Solín, Ľ., Analysis of floods occurrence in Slovakia in the period 1996–2006 (in Slovak). *Journal of Hydrology and Hydromechanics*, **56(2)**, pp. 95–115, 2008.

[11] Solín, Ľ., Cebecauer, T., Grešková, A. & Šúri, M., Small basins of Slovakia and their physical characteristics. *National Report 2000 of the IHP UNESCO Project 1.1.* FRIEND. Institute of Geography SAS, Slovak Committee for Hydrology. Bratislava, 2000.

[12] Solín, Ľ. Regional variability of flood threat in small basins of Slovakia (in Slovak*). Geographical Journal*, **63(1)**, pp. 29–52, 2011.

[13] Solín, Ľ. & Feranec, J., Land cover changes in small catchments in Slovakia during 1990–2006 and their effects on frequency of flood events. *Natural Hazards*, **56**, pp. 195–214, 2011.

[14] Solín, Ľ., Spatial variability in the flood vulnerability of urban areas in the headwater basins of Slovakia. *Journal of Flood Risk Management*, **5**, pp. 303–320, 2012.

[15] Solín, Ľ., Madajová, M. & Skubinčan, P. Analysis of commercial flood risk insurance in Slovakia. (Manuscript submitted for publication), 2013.
[16] Solín, Ľ., & Grešková, A., Small catchments of Slovakia: basic spatial units of hydrogeographical regional division (in Slovak). *Geographical Journal*, **51(1)**, pp. 77–96, 1999.
[17] Feranec. J., Otahel, J. & Pravda J., Land cover of Slovakia – identified by the CORINE land cover method (in Slovak). *Geographia Slovaca 11*. Institute of Geography, Slovak Aćademy of Sciences, Bratislava, 1996.
[18] FLOODsite, *Guidelines for Socio-economic Flood Damage Evaluation.* FLOODsite project Report number T9-06-01, 2006.
[19] Malczewski, J., *GIS and Multicriteria Decision Analysis*, John Wiley & Sons: New York, 1999.
[20] Malczewski, J., GIS-based multicriteria decision analysis: a survey of literature. *International Journal of Geographical Information Science*, **20(7)**, pp. 703–726, 2006.
[21] Weingartner, R., Barben, M. & Spreafico, M., Floods in mountain areas – an overview based on examples from Switzerland. *Journal of Hydrology,* **282**, pp. 10–24, 2003.
[22] Borden, A.K., Schmidtlein, M.C., Emrich, C.H.T., Piegorsch, W.W. & Cutter, S.L., Vulnerability of U.S. cities to environmental hazards. *Journal of Homeland Security and Emergency Management*, **4 (2)**, pp. 1–21, 2007.
[23] Cutter, S.L., Boruff, B.J. & Shirley, W.L., Social vulnerability to environmental hazards. *Social Science Quarterly*, **84 (2)**, pp. 242–261, 2003.
[24] Tapsell, S.M., Pennig-Rowsell, E.C., Tundtall, S.M. & Wilson, T.L., Vulnerability to flooding: health and social dimensions. *Philosophical Transactions of the Royal Society*, **360**, pp. 1511–1525, 2002.
[25] Simpson, D.M. & Human, R.J., Large-scale vulnerability assessments for natural hazards. *Natural Hazards*, **47**, pp. 143–155, 2008.
[26] Adger, N., Brooks, N., Bentham, G., Agnew, M. & Eriksen, S., New indicators of vulnerability and adaptive capacity. *Technical Report 7*, Tyndall Centre for Climate Change Research, Norwich, 2004.
[27] Damm, M., Fekete, A. & Bogardi, J.J., Intersectoral vulnerability indices as tools for framing risk mitigation measures and spatial planning. *Conference Proc. HydroPredict 2010*, Praha, CD ROM, 2010.

SESAME: improving small and medium enterprises' operational response and preparedness to flood events

G. Coates[1], M. McGuinness[2], N. G. Wright[3], D. Guan[4], T. Harries[5] & L. McEwen[6]
[1]*Durham University, UK*
[2]*University of Sheffield, UK*
[3]*De Montfort University, UK*
[4]*University of East Anglia, UK*
[5]*University of Kingston, UK*
[6]*University of the West of England, UK*

Abstract

Recent years have seen flooding emerge as a significant risk facing businesses in the United Kingdom (UK). This paper presents an overview of research aimed at improving small and medium enterprises' response to and preparedness for flood events, reducing the disruption and economic loss associated with flooding and strengthening the flood resilience of the wider economy and society. To achieve these aims, an understanding will be gained of business continuity processes and how small and medium enterprises (SMEs) behave in the immediate aftermath of flood events. Based on this understanding, agent based modelling and simulation will be used to investigate and assess how SMEs' behaviours during and in the short-term aftermath of flood events may influence their business continuity. Economic analyses will establish the local impacts of flooding in UK urban areas and the cascading effects to the wider economy. Iterations of agent based simulations will enable analyses of how changes in behaviour could influence business operational response and inform strategic decision making for long term flood preparedness. The research will inform guidance on how businesses may better adapt to flood risk through more effective business continuity planning, thereby increasing flood resilience.
Keywords: floods, decision making, businesses, response and preparedness.

1 Introduction

In recent years flooding has emerged as a significant endemic risk facing individuals, businesses and organisations across the UK. Indeed, the autumn and early winter of 2000–01 saw the worst flooding in the UK since 1947 and the 2007 floods resulted in the greatest loss of essential services since World War II. Flooding is characterised by high levels of uncertainty, unpredictability and emergence; key elements of hazards symptomatic of a 'risk society' [1, 2]. Hazards such as modern day flooding require novel response and regulation [3], and urgently, given that milder wetter winters and localised extreme weather events are likely to be an ongoing feature of both national and global climate patterns [4–6].

In 2012, the UK's Engineering and Physical Sciences Research Council funded three major projects focussing on innovative solutions to flood risk management. This paper presents an overview of one of those projects, called SESAME, which is aimed at influencing the behaviours of small and medium enterprises (SMEs) faced with flooding and flood risk. SESAME aims to enable businesses of this nature to enhance their capabilities for both short term event response and longer term preparedness. Thus, SMEs will be able to reduce the disruption and economic losses associated with flooding and, as a result, strengthen the flood resilience of both the wider economy and society. To achieve this aim, four integrated and interdisciplinary research goals are being pursued. Firstly, there is a need to achieve a better understanding of business continuity processes within SMEs and how these businesses behave and interact with each other in the immediate and longer term aftermath of flood events. Secondly, there is a need to establish how agent based modelling and simulation can be used to represent key attributes and simulate behaviours of at-risk UK small businesses and investigate the consequences of these for both the short-term response to flooding and longer-term preparedness. Thirdly, there is a need to assess the impacts of flooding on economic systems both within and beyond the immediately affected area and explore how changes in business behaviour could influence these impacts. Fourthly, there is a need to develop innovative approaches to the promotion of organisational behaviour change and adaptive organisational learning throughout the flood cycle. In pursuing each of these four goals, the research exploits synergies between the academic fields of organisational business continuity management, agent based modelling and simulation, disaster management, flood modelling and prediction, economic modelling and the social/behavioural sciences.

2 Background

It has been estimated that flood damage could account for £10bn to the UK economy in 25 years [7] and that risks relating to precipitation over the next 30 to 100 years will increase by 2 to 4 times across the UK, with some locations experiencing changes beyond this range [8]. A key recommendation emerging from the formal review of the 2007 UK floods was the need to reduce the risk of flooding and its impact [5]. The UK's National Security Strategy [9] discusses the importance of flexibility and resilience to natural and man-made disasters, but also emphasises working in alliances and partnerships to build corporate and community resilience. This can be best understood as part of a paradigm shift in approaches to the management of risk itself. Policy developments in the UK which address risk and resilience are increasingly underpinned by concepts such as communal risk reduction and also, through the involvement of a range of non-state bodies, risk spreading [10]. One powerful stakeholder is the insurance industry. The cost to insurers of the 2007 UK floods alone was £3bn with insurers highlighting that losses on this scale are not sustainable for the insurance industry. Currently, under a pan-industry agreement with government, British insurers are obliged to make insurance cover for flooding as widely available as possible and, in effect, insure properties that would otherwise be deemed uninsurable [11]. In the future, it is probable that insurance coverage for flood risk will become less available and more expensive. The impact of this on businesses, and particularly SMEs, is significant given that SMEs account for over 50% of UK employment and gross domestic product. However, in light of the paradigm shift mentioned and in the context of the current fiscal crisis, it is unlikely that government will wholly absorb the risk and associated costs. Rather, it is likely that there will be an even greater emphasis on risk spreading as a means for building resilient cities and communities. Citizens, private and public sector businesses, organisations and bodies will increasingly be obliged to engage in the management of flood risk and the creation of resilience at the local community, regional and national level. Indeed, business, organisational and local resilience are vital interacting components of an overarching strategy for national resilience.

3 Related work

Research on the response to major natural and man-made events has focussed primarily on 'category 1' responders [12]. However, there is a scarcity of research on the response of businesses faced with major disruption caused by flooding. Effective response to disruption through preparedness can mitigate the impact of a major flood event on business organisations. This approach forms a cornerstone of business continuity management [13], which is recognised as an important tool for organisational survival in the face of disruption [14, 15]. Effective business continuity can reduce organisational down-time and facilitate a more rapid return to normality thereby lessening the economic impact at the level of the organisation and beyond. Previous reviews have highlighted the role of business continuity as part of any successful flood response [5]. Despite this, in the UK, organisational engagement with business continuity remains low with less take-up by SMEs and third sector organisations relative to larger businesses and public sector bodies [16, 17]. There are indications that a lack of resources and capabilities may account for this lower take-up, however no systematic investigation has yet been undertaken.

Whilst agent based models (ABMs) have provided useful insights into emergency responses to major events [18, 19], their focus to date has been on agents representing individuals exhibiting behaviour intended to mimic members of the public or emergency responders. In flood management, ABMs have been used to model evacuation strategies, albeit without agents representing emergency service personnel [20, 21]. In the response to flood events, there exists scope to develop and utilise ABMs by representing and modelling businesses, rather than individuals, as agents and considering timeframes beyond immediately after the event occurs. ABMs offer an opportunity for using and embedding stakeholder learning and feedback within the modelling process. This can occur through engagement with model outputs from agent based simulations of past or future flood events based on historical data and/or predictions from flood modelling. Indeed, predictions from flood modelling can complement incomplete observed data for specific events [22, 23], postulate future scenarios based on organisation agent behaviour and provide flooding frequencies [24]. Feedback on behaviour changes, gained from working with stakeholder businesses and organisations, can then be fed back into further simulations of the ABM, so improving modelling of the impact of changes in businesses' preparation for and response to flood hazards.

4 The SESAME project

The SESAME project consists of a number of integrated strands of research (as indicated in fig. 1). SESAME aims to iteratively model and evaluate businesses', in particular SMEs', response strategies to flooding in order to develop guidance to influence behaviour change thus enabling them to better prepare for future floods and adapt to flood risk through more effective continuity planning.

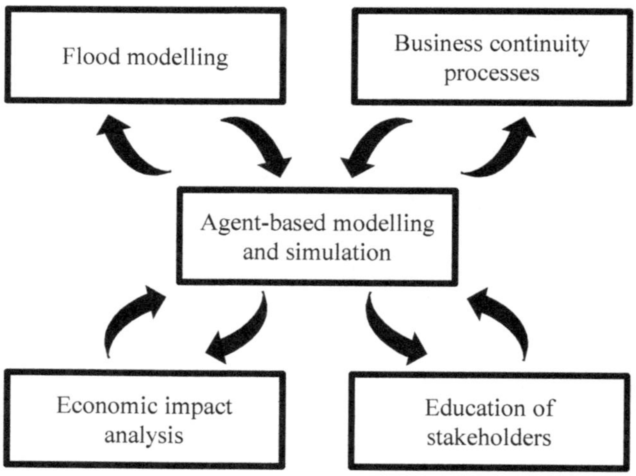

Figure 1: Integrated research stands of the SESAME project.

This aim will be brought about through the integrated application of flood modelling with agent based modelling and simulation of SMEs informed by business continuity processes, analysis of the economic impacts radiating out from directly flood-affected areas and the education of stakeholders in better preparing for flood events.

4.1 Flood modelling and SME level agent based modelling and simulation

The focus of this strand of the research is to develop an agent based approach to model and simulate SME behaviour in the face of flood events [25]. This involves the representation of SME's attributes and behaviour in terms of their actions and interactions with other organisations in response to flood events, thus informing strategic decision making for longer term preparedness.

An overview of the computer modelling and simulation work being undertaken in the SESAME project is presented in fig. 2. Stage 1 involves developing the agent-based model's virtual geographic environment, which is able to combine layers of Ordnance Survey (OS) information with flood model output, in Stage 2, in order to identify the businesses affected by flooding. Flood model output is dynamic in that the flood inundation varies with time thus bringing a temporal aspect to simulations performed in Stage 5. Stage 3 relates to developing business agents to model SMEs and other types of organisations, in terms of their attributes, behaviours, actions and interactions in response to flood events. Stage 4 involves setting-up an agent interaction network to enable simulations to be performed, in stage 5, thus informing SMEs how they might change their behaviour to better prepare for and respond to future flood events.

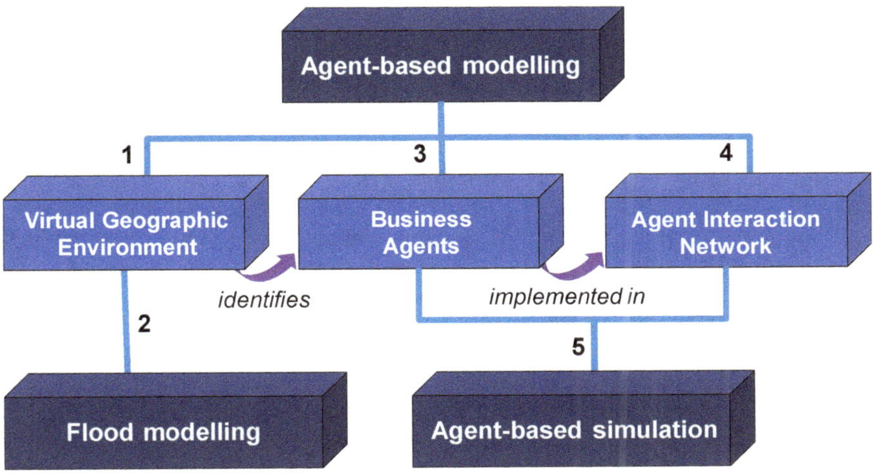

Figure 2: Computer modelling and simulation.

The virtual geographical environment (VGE) component of the ABM [26], developed using OS MasterMap, is fed information regarding a dynamic inundation prediction thus allowing flood events to be modelled in any urban area of the UK. Consequently, all businesses and organisations, including SMEs, in this area directly and indirectly affected by a flood event can be identified. An example VGE modelling an area affected by a flood event is shown in fig. 3.

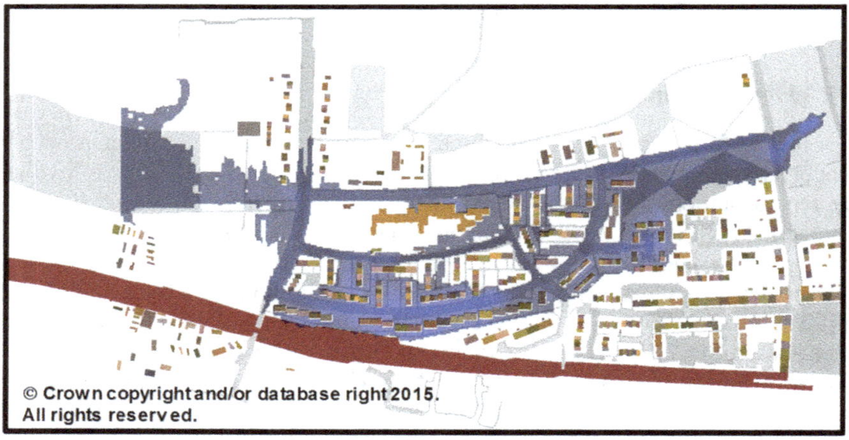

Figure 3: Example VGE modelling a geographical area affected by a flood event.

Within the affected geographical area, intelligent agents are used to model SMEs with attributes and rules governing their operational behaviour in terms of actions and interactions with other organisations including businesses. In addition to the VGE illustrated in fig. 3, an agent interaction network will capture, in simulations of floods events and the aftermath, the interdependent behavioural actions of businesses and other organisations. Multiple simulation experiments will be designed and carried out reflecting SMEs' different types of behaviours. This will be done in two stages to gain an understanding of the dynamic complexity of businesses' interactions in the aftermath of a flood: (1) simulations in the context existing behaviours exhibited by SMEs according to semi-structured interviews with SMEs which have experience of significant flooding; (2) simulations with changes in SME behaviour adopted in light of the simulations in (1) and other sources including guidelines from the Environment Agency and local councils, business continuity management systems' requirements (ISO223301) and academic literature.

4.2 Business continuity processes

This strand of the research is aimed at gaining an understanding of SMEs' response to flood events in the short and longer term in order to generate insights and data regarding the attributes, dynamic behaviour and interactions of such businesses. These insights and data will form the basis of the attributes and rules governing the operational behaviour of SMEs represented in the ABM referred to in Section 4.1. To achieve this aim, this research strand will explore the factors which underpin both effective and ineffective business response to flooding by identifying what, why and when SMEs behave in particular ways in the flood response and preparedness phases. This will involve examining how long it takes these businesses to move through the response phase (short term) and preparedness phase (longer term) and regain normal operational functioning so that a better understanding can be gained of what constitutes short term and longer term with respect to effective flood risk. SMEs' behaviours are influenced directly and indirectly through interactions with a range of other organisations and stakeholders such as quasi-governmental agencies, insurers and other businesses, for example across supply and distribution chains. Consequently, it is not possible to generate meaningful insights into these behaviours without contextualizing these actions dynamically within this broader network of actors/agents. By taking this holistic perspective, it is possible to identify what barriers may exist to effective business continuity and where these barriers lie, for example at the micro or meta level. To generate robust and rich data, a multi-method approach is being taken through integrating traditional methods of data collections, such as interviews and focus groups, with data captured from new technology for a number of flood event case studies. Importantly, case studies being considered vary in terms of severity as well as flooding source such as pluvial, fluvial and coastal. As mentioned earlier, data generated will feed into the ABM to enable the simulation of business behaviour. Subsequent changes in actions and behaviours will then be interrogated iteratively, as well as directly with stakeholders through appropriate means such as follow-up interviews and online forums, blogs and wikis.

4.3 Impact of flooding to urban economy and wider economic systems

The emphasis of this research strand is the economic analysis of the cascading impact of flood events, which is central to identifying economic vulnerabilities and thus where preparedness measures are best applied. Assessments of the economic impact of flood events in urban areas have traditionally focused locally on the initial estimates on the cost of lost assets, referred to as direct damage. These initial estimates are useful in establishing where capital and resources are needed for rebuilding post-event. However, since economies are interlinked geographically, there is a need to establish how this direct damage propagates from the local economy throughout the wider economic systems at regional and national levels, thus determining the indirect damage. Quantified knowledge on flooding impact will be compiled, including the cost of direct damage and business interruption based on DEFRA's Multi-coloured Manual, into a set of damage functions at an urban area scale. A variant of the adaptive regional input-output (ARIO) model will then be developed to explore the vulnerabilities of urban area economies related to flood damage. This model enables direct damage to be determined using an event accounting matrix, consisting of a set of damage functions. The indirect damage is determined by integrating the local economic datasets and direct damage datasets based on supply constraint environmental input-output analysis. Using the variant ARIO model, economic analyses can be performed regarding the impacts of flooding in UK urban areas leading to specific economic vulnerabilities being identified, which should be the focus of increased preparedness. That is, the identification of business sectors beyond the immediate, or local, impact zone of a flood which are key to the wider economy at a regional and national level.

4.4 Adaptive e-learning and behaviour change in organisations through the flood cycle

This strand of research addresses the need to develop new ways of bridging knowledge gaps between outputs from agent based simulations and the promotion of real and timely behaviour change for resilience among the different stakeholders involved in business contingency planning for floods. As such, a key aspect of this strand is engagement with stakeholders such as businesses, local authorities and the Environment Agency. The development of the ABM mentioned earlier presents opportunities for improving the interface with a broad range of end-users and the tailoring of model outputs to the needs of diverse organisations. Feedback on iterative organisational learning and predicted behavioural changes, after using model outputs from single and/or multiple flood simulations as practical guidance, can then be embedded into the ongoing ABM design process and future simulations. This will improve the underlying assumptions in the modelling that determine the outputs. This research strand is also aimed at developing an e-learning intervention tool that uses new digital technologies to engage end-users with the practical guidance established, and evaluates their use for learning and behavioural change. As smart, future-oriented ubiquitous technologies will be dominant modes of future information engagement, design principles developed will be applicable across diverse flood modelling settings.

5 Conclusions and future work

It is widely acknowledged that the effect of future floods in the UK is expected to be significant in terms of damage, disruption and economic loss. Thus, the need to reduce the risk of flooding and its impact has been identified as key to mitigating this effect. The overview presented in this paper relates to an ongoing research project, termed SESAME, aimed at addressing the need to reduce the impact of flooding by improving SMEs' response to and preparedness for flood events, reducing the disruption and economic loss associated with flooding and strengthening the flood resilience of the wider economy and society.

Agent based modelling and simulation coupled with flood modelling provides a means of (i) identifying businesses affected by a flood event in a geographical area, (ii) investigating the actions and interactions of SMEs with other organisations, in terms of their response and preparedness to flood events, and (iii) evaluating the implications of strategic decisions of SMEs in the context of changes in their behaviour. Key to such an agent based approach is the need to determine and codify how businesses respond to and prepare for flooding at an operational level. Engagement with a variety of SMEs will establish the business continuity processes and dynamic behaviours of these businesses in flood response and preparedness thus providing data inputs to an ABM. Subsequently, agent based simulations will identify key drivers and barriers to enhance SME resilience through continuity planning. Further, simulations will inform guidance regarding enhancements and adaptation in all stakeholders' response and preparedness in the face of flood events, which will be communicated to relevant businesses via relevant risk communication and pedagogical concepts.

Acknowledgement

The authors gratefully acknowledge the funding provided be the UK's EPSRC under grant EP/K012770/1.

References

[1] Beck, U., *Risk Society: Towards a New Modernity*, Sage Publications Ltd: London, 1992.

[2] Beck, U., *World at Risk*, Polity Press: Cambridge, 2009.

[3] Hood, C., James, O. & Scott, C., Regulation of government: has it increased, is it increasing, should it be diminished? *Public Administration*, **78(2)**, pp. 283-304, 2000.

[4] Stern, N., Stern Review on the Economics of Climate Change, HM Treasury /Cabinet Office, 2006.

[5] Pitt, M., The Pitt Review: Lessons learned from the 2007 floods, Cabinet Office, 2008.

[6] UK Climate Impacts Programme (UKCIP), *A changing climate for business*, UKCIP: Oxford, 2010.

[7] The Confederation of British Industry (CBI), *Future Proof: preparing your business for a changing climate*, CBI: London, 2009.

[8] Foresight: Future Flooding, *Foresight: Future Flooding Study, Flood and Coastal Defence Project: Office of Science and Technology*, 2004.

[9] HM Government, A Strong Britain in an Age of Uncertainty: The National Security Strategy, HM Government, 2010.

[10] Walker, C. & Broderick, J., *The Civil Contingencies Act 2004: Risk, Resilience and the Law in the United Kingdom*, Oxford University Press, 2006.

[11] Institution of Civil Engineers (ICE), *Flooding: Engineering resilience*, ICE: London, 2008.

[12] Civil Contingencies Act 2004, *Act of Parliament of the United Kingdom*, 2004.

[13] British Standards Institution (BSI), *ISO 22301 Business Continuity Management*, BSI: London, 2012.

[14] Elliott, D., Herbane, B. & Swartz, E., *Business Continuity Management*, Routledge: London, 2001.

[15] Herbane, B., The evolution of business continuity management: A historical review of practices and drivers. *Business History*, **52(6)**, pp. 978-1002, 2010.

[16] Musgrave, B. & Woodman, P., *Weathering the Storm: The 2013 Business Continuity Management Survey*, Chartered Management Institute: London, 2013.

[17] Herbane, B., Small business research: Time for a crisis-based view. *International Small Business Journal*, **28(1)**, pp. 43-64, 2010.
[18] Bellamine-Ben Saoud, N., Ben Mena, T., Dugdale, J., Pavard, B. & Ben Ahmed, M., Assessing large scale emergency rescue plans: an agent based approach. *International Journal of Intelligent Control and Systems*: Special Issue on Emergency Management Systems, **11(4)**, pp. 260-271, 2006.
[19] Hawe, G.I., Wilson, D.T., Coates, G. & Crouch, R.S., Investigating the effect of overtriage on hospital arrival times of critically injured casualties during a major incident using agent-based simulation. *Proc. of the 6th International Conference on Soft Computing and Intelligent Systems and the 13th International Symposium on Advanced Intelligent Systems*, Kobe, Japan, 2012.
[20] Liu, Y., Okada, N., Shen, D. & Li, S., Agent based flood evacuation simulation of life-threatening conditions using Vitae system model. *Journal of Natural Disaster Science*, **31(2)**, pp. 33-41, 2009.
[21] Dawson, R., Peppe, R. & Wang, M., An agent based model for risk-based flood incident management. *Natural Hazards*, **59(1)**, pp. 167-189, 2011.
[22] Hunter, N.M., Bates, P.D., Neelz, S., Pender, G., Villanueva, I., Wright, N.G., Liang, D., Falconer, R.A., Lin, B., Waller, S., Crossley, A.J. & Mason, D.C., Benchmarking 2D hydraulic models for urban flood simulations. *Proceedings of the Institution of Civil Engineers: Water Management*, **161(1)**, pp. 13-30, 2008.
[23] Neal, J.C., Bates, P.D., Fewtrell, T.J., Hunter, N.M., Wilson, M.D. & Horritt, M.S., Distributed whole city water level measurements from the Carlisle 2005 urban flood event and comparison with hydraulic model simulations. *Journal of Hydrology*, **368**, pp. 42-55, 2009.
[24] McMillan, H.K. & Brasington, J., End-to-end flood risk assessment: A coupled model cascade with uncertainty estimation. *Water Resources Research*, **44(3)**, W03419, 2008.
[25] Li, C. & Coates, G., Design decisions in the development of an agent-based model for business operations when faced with flood disruption, *Proceedings of the International Conference on Complex Systems in Business, Administration, Science and Engineering*, New Forest, England, 2015.
[26] Coates, G., Hawe, G.I., Wright, N.G. & Ahilan, S., Agent-based modelling and inundation prediction to enable the identification of businesses affected by flooding. *Flood Recovery, Innovation and Response IV, WIT Transactions on Ecology and the Environment*, **184**, pp. 13-22, 2014.

Analysis of the loss probability relation on a community level: a contribution to a comprehensive flood risk assessment

M. Huttenlau[1], K. Schneeberger[1,2], B. Winter[1,2], J. Reiss[1,2] & J. Stötter[1,2]
[1]*alpS, Centre for Climate Change Adaptation, Austria*
[2]*Institute of Geography, University of Innsbruck, Austria*

Abstract

Flood risk management is an appropriate method for understanding and mitigating the consequences of flooding which regularly causes significant losses to property and human lives. Therefore, the analysis of flood risk is a key issue. Within this risk analysis, direct tangible damages are important flood hazard impact indicators. This paper shows an approach of how to assess the potential direct monetary damages in a mountain meso-scale study area (Austrian Province of Vorarlberg). The approach comprises the following key components: (i) asset pooling and its monetary assessment, (ii) exposure analysis and (iii) susceptibility analysis. By applying this method, loss probability relations reflecting the potential direct monetary damages of flood events with a return period of 30 yrs, 100 yrs and 300 yrs were derived for all flood exposed communities in the study area. The results illustrate the possible uncertainties that are inherent in the procedure of assessing direct monetary damages. Highlighting this significant variability is a valuable contribution to a comprehensive flood risk assessment framework.
Keywords: flood risk, damage functions, asset assessment, direct tangible damages.

1 Introduction and background

Flood risk analyses are valuable instruments for decision making in risk management. Knowledge about the costs associated with flood events is important for a wide range of application fields, e.g. cost-benefit considerations in planning procedures like river training, premium pricing in the insurance industry, calculation of reinsurance coverage or political decision making (e.g. adequate budget of disaster relief funds).

The general risk concept is based on the understanding of consequences (mostly losses) with a given probability whereby numerous definitions of risk exist [1, 2]. In this study, the definition of the European Flood Directive is followed which defines flood risk as "… the combination of the probability of a flood event and the potential adverse consequences for human health, the environment, cultural heritage and economic activity associated with a flood event" [3]. In order to reflect the flood risk situation in a study area adequately, flood risk analysis should (i) consider the whole range of possible scenarios; (ii) relate the scenarios to impact consequences and (iii) carry out a probabilistic risk assessment [4–6].

In the framework of most flood risk analyses, homogenous hazard scenarios with a constant return period (e.g. RP: 100 yrs) throughout an entire river basin are considered. As stated by Thieken *et al.* [5], the assumption of a constant return period along the relevant river reach or throughout an entire catchment must be questioned, especially for large catchments under study on meso- and macro-scale. To overcome this methodological deficiency, the generation of artificial flood hazard scenarios in combination with flood impact analysis seems to be a promising approach. Initial studies applying multivariate distribution functions to generate heterogeneous scenarios were recently published [5, 7, 8].

The Austrian Federal Province of Vorarlberg was affected by serious flood events (e.g. 1999, 2002 and 2005) in recent decades and therefore a thorough flood risk analysis was urgently needed. In strong collaboration with a regional insurance company, a probabilistic flood risk assessment framework was developed for Vorarlberg. This approach has been applied to assess the flood risk of specific insurance portfolio and of the entire assets in the study area. As a conclusion from the discussed drawbacks of using a homogenous scenario in flood risk analysis, this approach is based on heterogeneous scenarios. The framework consists of three major components: (i) a comprehensive flood data catalogue as a surrogate for the analysis of the probability of flooding, (ii) potential flood damages (e.g. direct monetary damages) and (iii) a probabilistic flood risk assessment procedure [9].

Probabilistic risk assessment depends on a large number of flood events/scenarios. Since the number of past damaging flood events is usually relatively small, further events have to be produced artificially to obtain a comprehensive flood data catalogue. Such events are generated by applying the semi-parametric conditional approach proposed by Heffernan and Tawn [10] which is based on time series derived from river gauging stations. As the approach, so far, only comprises information on point scale, there is a need for spatial interpolation of all flood events within the comprehensive flood data catalogue over the entire river network. These heterogeneous scenarios comprise information about the flood return period along the entire river network. The individual scenarios are combined with loss probability relations, which reflect the results of a flood impact study. Subsequently, a synthetic time series of flood losses is generated and statistically evaluated [9].

The presented work focuses on the determination of loss probability relations of communities which are exposed to river floods. The methodology is derived from the frequently applied risk analysis procedure in natural hazard research. The central idea is the application of damage functions to derive the direct monetary flood damages in a study area. In general, the approach of the presented study follows the recommendations of Merz *et al.* [11] with the main difference being the reordering of the individual steps. Further variations concerning the general methodological framework and the structure of risk-based analysis can be found in the scientific literature [12–14]. All the methodological variations and related studies aim at analysing natural hazard induced consequences assuming a given occurrence probability. In order to achieve this aim, hazard scenarios are systematically combined with socio-economic values. Finally, the consequences of the hazard process on the values are analysed.

This contribution concentrates on the assessment of the direct monetary damages at a community level as a major component of the briefly introduced framework for a probabilistic flood risk assessment. As the study has a distinct actuarial background, the contribution focuses on direct tangible damages to buildings and contents (or inventory) based on replacement values. In this contribution, specific attention is given to the damage assessment part of the introduced framework. In this manner, the problem of the damage assessment (compared to the hazard assessment) within risk analysis not often being well balanced (as also emphasized by Merz *et al.* [11]) is practically overcome.

2 Study area and its flood characteristics

The study area covers the entire Federal Province of Vorarlberg, Austria with a total area of ca. 2600 km^2, of which 21.8% is defined as area of permanent settlement (area suitable for building use, transport and agriculture) [15]. Located in the Eastern Alps, Vorarlberg is characterized by a complex mountain topography ranging from 395 m.a.s.l. (Lake Constance) up to 3312 m.a.s.l. (summit of Piz Buin). The main settlement area concentrates on the Rhine Valley (Feldkirch to Bregenz, concentration of two thirds of the population) and its tributary valley floors, like the valley of the river Ill (Walgau) or the valley of the Bregenzer Ache (fig. 1). Approximately 380,000 people are living in Vorarlberg corresponding to an average population density of 145 inhabitants per km^2.

From a hydrological perspective, the major part of the investigated region belongs to the Alpine Rhine basin. The study area can be subdivided into three parts: the flat and densely populated Rhine-Valley, the southern catchment of the river Ill and the catchment of the Bregenzer Ache, a direct tributary to Lake Constance [16].

In the last decades, severe flood events have regularly affected communities in Vorarlberg. As an example, the well documented flood event in 2005 was triggered by intensive rainfall over 30 hours with precipitation rates up to 10 mm/h. Due to high soilwater saturation and the zero degree limit above

2900 m.a.s.l., a high proportion of the precipitation had immediate runoff effects. The overall direct damage in Vorarlberg, covering infrastructure, industries and private property, was estimated to ca. € 180 million [17].

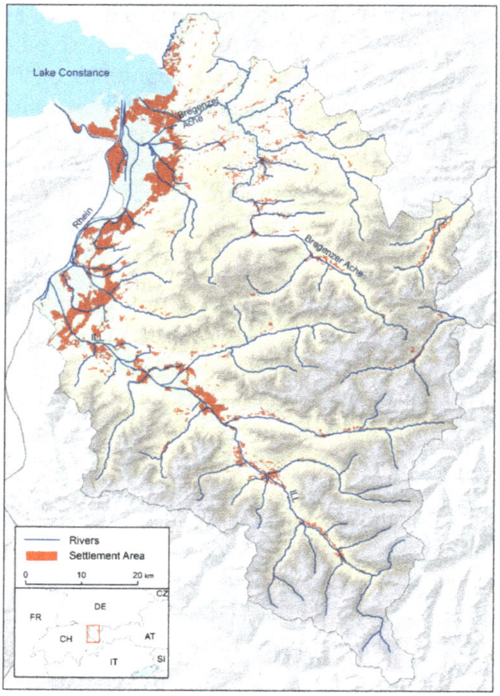

Figure 1: Study area with river network and settlement areas.

3 Data and methods

The methodological framework of this assessment of direct monetary flood damages follows the procedure according to Merz *et al.* [11] with some differences regarding the reordering of individual steps. The assessment comprises the following three steps:

1. Preprocessing: (i) definition and editing of hazard scenarios, and (ii) asset pooling and monetary asset assessment.
2. Exposure analysis: identification of the elements at risk within certain hazard scenarios and assignment of the corresponding hazard impact.
3. Susceptibility analysis: analysis of expected damages by relating damage functions with information of the exposure analysis regarding the elements at risk.

In research, the third step is often referred to as vulnerability analysis. However, the term 'vulnerability' is used in many risk approaches/concepts of various disciplines (e.g. engineering, geography, sociology or economy); as a consequence, a commonly accepted definition does not exist [12, 14]. Thus, the authors follow the suggestions of Merz *et al.* [11] and apply susceptibility analysis as a last step to calculate expected damages.

3.1 Preprocessing

The relevant geodata were made available by public authorities (e.g. land surveying office, Vorarlberg). The applied address dataset was made available by the Austrian Post (Österreichische Post AG), all insurance data were made available by Vorarlberger Versicherung VaG.

3.1.1 Hazard scenarios

The severe flood event in August 2002 was a trigger for flood risk related research activities in Austria, amongst others the initiation of the national flood mapping project 'HORA' (HOchwasser Risikoflächen Austria – flood risk zones in Austria) [18]. One of the objectives of the HORA project was the generation of inundation maps and their public access as it is also required by the European Flood Directive [3]. A digital elevation model (DEM) with a spatial resolution of 10 m provided the basis input for (i) the 1-D hydraulic simulation and (ii) the final inundation maps. As the uncertainties associated with this flood maps were considerable, a further development of the model was necessary [19]. Within this improved setup, a DEM with a 1 m resolution was applied. Additionally, major river reaches with high loss potential were simulated with a 2-D hydraulic simulation. Thus, official inundation maps of good quality approved by a federal authority are available for a return period of 30 yrs, 100 yrs and 300 yrs for the entire study area.

3.1.2 Asset pooling and monetary asset assessment

The study focuses on direct tangible damages to buildings and contents based on replacement values. In a previous research [6, 20], it was shown that portfolio information and derived insured amounts from insurance companies with a high market share in the investigated study area can be very valuable to assess the assets. In contrast to this previous study with a focus on all individual properties of the entire study area, the present study additionally considers elevation of buildings based on high resolution digital terrain and digital surface models (spatial resolution of 50 cm). Thus, mean values of buildings based on cross cubature values can be statistically analysed under consideration of the available insurance portfolio and further extrapolated to the entire building stock.

Therefore, a geodatabase was established which consists of geocoded addresses, building footprints, official land destination plans and building object categories, high resolution aerial images, digital terrain and digital surface models, classified LiDAR point cloud (to derive building footprints) and further data. As a first step, the available insurance portfolio was geocoded (matching rate above 90%; the rest lacking correct spatial information). The building object categories were combined with the address data and aggregated to six categories (private residential, hospitality industry, office buildings, wholesale and retail sector, industry and other objects), for which mean values of the insured buildings were statistically analysed on a district level: (i) building cross cubature values for the individual buildings based on the available data (statistical analysis of building footprint and calculated normalized digital surface model) and (ii) mean insured amount per household or content unit. Finally, the derived mean values (indexed to 2013) were extrapolated to the entire building stock in the study area.

3.2 Exposure analysis

By means of the exposure analysis, the objects which are affected by the hazard process (i.e. flood scenarios) are identified. The hazard scenarios are represented by the inundation maps with a return period of 30 yrs, 100 yrs and 300 yrs. All elements (object based) in the study area were intersected with inundation data. Therefore, the elements at risk associated with the three inundation scenarios were identified including the corresponding water depths.

3.3 Susceptibility analysis

The final step to estimate the direct monetary damages is the susceptibility analysis. The concept of damage functions (or susceptibility functions) can be regarded as the central idea, where these functions express the susceptibility of the certain types of objects at risk. The damage functions describe the relation between one or more hazard parameters and the relative damage from which the monetary damage can be calculated [11, 21].

The most frequently used hazard parameter is water depth, although various other factors, such as flow velocity, duration of inundation, sediment concentration and contamination of flood water are also relevant [11]. According to the number of influencing parameters, damage functions can be differentiated into single- and multi-parameter models. Besides impact parameters (e.g. water depth), multi-parameter models often incorporate resistance parameters. Whereas impact parameters depend on the hazard process, resistance parameters are mostly related to the properties of the elements at risk [22]. As they are heterogeneous in space and time, difficult to predict and limited information exists about their effects, the additional factors of multi-parameter models are hardly transferable and thus they are often neglected [11]. Within this study, water depth is the only considered damage parameter.

Besides the damage parameter, the susceptibility functions comprise the different categories of the elements at risk. Damage functions for different object classes based on one methodological development can be referred to as damage model. The flood damage models can be differentiated into empirical and synthetic approaches. Empirical models are derived from damage data collections of ex post event data and synthetic approaches which consider ‘what-if-scenarios’ [11]. A further distinction into relative and absolute damage function is possible and its implication on a comparative study has to be considered. Table 1 gives an overview of the considered damage models in this study.

Table 1: Damage models and key characteristics.

Name	Origin	Object categories	Content	Example/reference
BUWAL	CH	7	separate	[23, 24]
Damage scanner	NL	8	included	[25, 26]
Flemish Model	BE	7	separate	[26]
FLEMO	DE	8	separate	[27, 28]
Hübl and Kraus	AT	1	included	[24, 29]
IKSE	DE	5	included	[30]
JRC	EU	5	separate	[26]
MURL	DE	7	separate	[31]
Rhine Atlas	DE	7	separate	[32]
Romang	CH	1	included	[24, 33]
Schmidtke	DE	4	included	[34]

4 Results

A procedure for the assessment of tangible direct monetary damages of buildings and contents was applied in the study area Vorarlberg. The loss probability relation for all flood-exposed communities (64 were affected) were calculated considering eleven different damage models (table 1). These loss probability relations illustrate the maximum potential damages for a return period of 30 yrs, 100 yrs and 300 yrs on community level (fig. 2). An accumulation of the potential damage of the affected communities leads to a maximum potential direct damage in the study area (fig. 3).

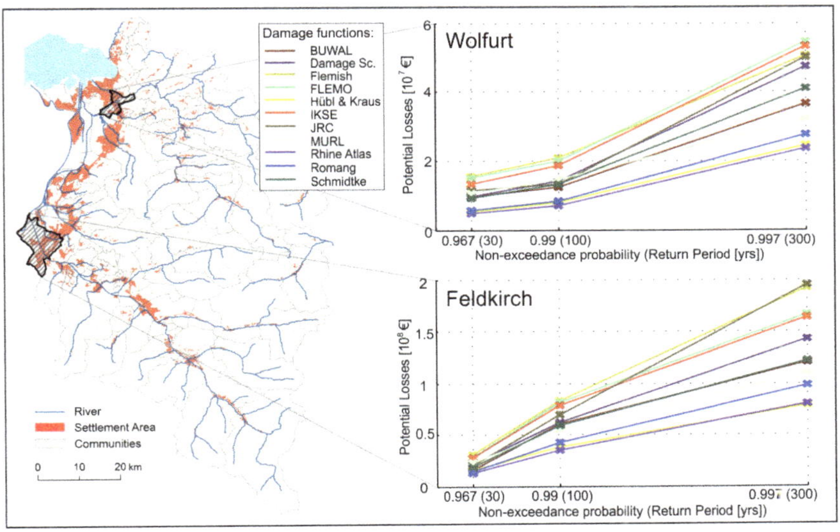

Figure 2: Loss probability relation of two communities and allocation within the study area.

The results of the different damage models differ significantly; for instance the potential damage of the damage model which leads to the highest losses (i.e. FLEMO) is 175% higher compared to the lowest one (i.e. Rhine Atlas) (RP: 100 yrs). Fig. 3 also shows that the potential damage of the model which leads to the highest losses for a RP of 30 yrs is almost the same as the losses associated with the lowest model for a RP of 300 yrs. The loss probability relations reflect the maximum possible direct damages associated with different damage models which can occur (triggered by river floods) in each community (exemplarily shown in fig. 2) and the entire study area (fig. 3), respectively.

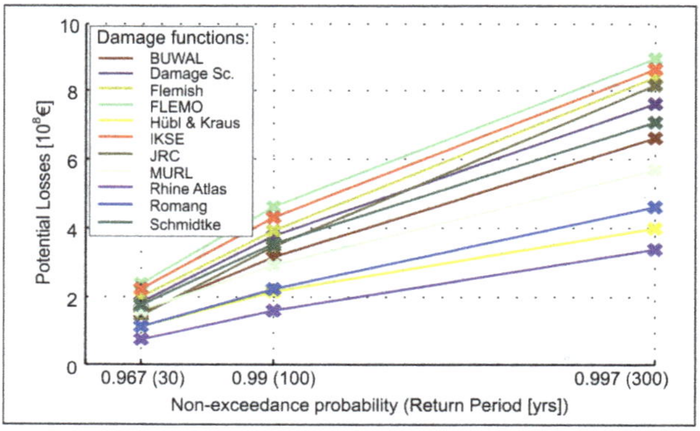

Figure 3: Accumulated loss probability relation of Vorarlberg.

5 Discussion and outlook

The assessment of the tangible direct monetary flood damage as a possible flood hazard impact indicator poses an essential component within a probabilistic flood risk assessment framework. As the loss probability relations have a linear effect on the final framework results, the consideration of one single damage model might lead to significant under- or overestimation of the monetary flood risk. In order to reveal the possible uncertainties, the procedure to assess the direct monetary damages has to be thoroughly examined. The following components are most important:

- Hazard scenarios represented by the flood inundation maps which rely on consistent information regarding the floodplain geometry are of high importance [35].
- Uncertainties related to asset pooling and monetary asset assessment are linked to two key issues. First, pooling has to match the different categories of all damage functions, which leads to best match assumptions. Second, extrapolation of mean values to the entire building stock is most suitable for the large number of residential buildings, while for most heterogeneous categories such as trade and industry the results lack a sound statistical basis.
- As the exposure analysis is based on single objects, the associated uncertainties are relatively small. Uncertainties arise mainly from the

inundation maps as described above. The potential elements at risk (object based) are accurately represented within a geographical information system and thus, compared to widely used area based approaches [26], the uncertainties are small.

- In this study a susceptibility analysis was conducted for eleven damage models. To enable a comparison of different damage models, a few restrictions have to be imposed: (i) absolute damage models were not considered and (ii) the water depth was the only damage impact parameter. Fig. 3 shows that the influence of the different damage models on the loss probability relation are considerable. The loss probability relations at a community level are linearly interpolated among the three concretely modelled return periods (RP: 30 yrs, 100 yrs and 300 yrs). This simplification does not take into account the possible rapid rise on certain water level thresholds (e.g. due to overtopping of embankments). Finally, it has to be questioned whether damage models are transferable in space and time [21].

In general, risk analyses are always associated with considerable uncertainties, which result from the several input parameters involved in the risk analyses procedure and the model itself. It is a common procedure in hydrology (and in other disciplines) to ensure a minimum level of quality by comparison of observed and simulated data. This kind of validation would also be desirable in flood risk analysis but, as extreme flood events and their adverse consequences can only be seldom observed so far, a traditional validation is hardly possible [5]. Within this study, the monetary damage of eleven different damage models was examined, with the results differing by a factor of 3 (see fig. 3). A comparison (i.e. plausibilisation) of observed loss data of a local insurance company and of ex post analysis data from public authorities with the model result of the reconstructed flood event 2005 (fig. 4) indicates that the low to moderate damage models deliver the most appropriate results in the study area of Vorarlberg.

The loss probability relations provide information about the maximum possible direct monetary damages corresponding to certain recurrence intervals at community level and for the entire study area. As homogenous flood scenarios (i.e. widespread flooding with constant RP) are very unlikely, a probabilistic flood risk assessment framework which considers a wide range of possible flood events is necessary to gather realistic results. Therefore, on the one hand a comprehensive data catalogue of flood events is a fundamental surrogate for the probability of flooding and on the other hand, the loss probability relations on community level provide information regarding the potential flood damage. These two components can be combined and the resulting time series of flood losses provides the fundamentals for statistical flood risk analysis [9]. Consequently, the loss probability relations are a valuable contribution to the framework of probabilistic flood risk assessment, which is an important feature in the flood risk management cycle.

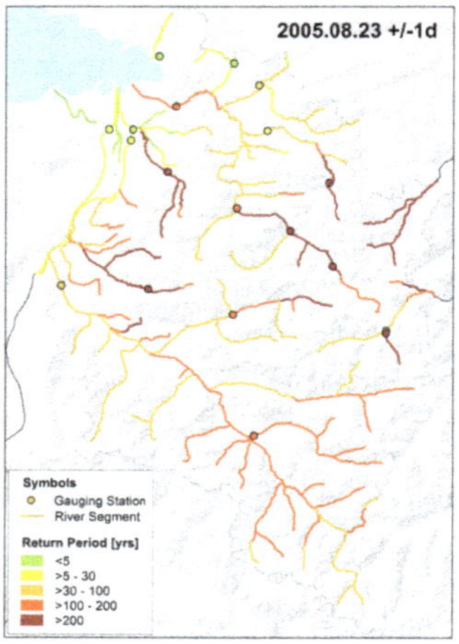

Figure 4: Reconstructed flood event from August 2005 [9].

Acknowledgements

This work results from the research project *InsuRE* funded by the Austrian Research Promotion Agency (FFG) within the scope of the programme COMET, the Tyrolean Insurance Company (Tiroler Versicherung), the Vorarlberger Insurance Company (Vorarlberger Landesversicherung). We would like to thank all the institutions that provided data, particularly the Hydrologic Offices and Austrian Post. K.S. acknowledges the financial support provided through the scholarship 'Nachwuchsförderung der Universität Innsbruck'.

References

[1] Kaplan, S. & Garrick, B.J., On the quantitative definition of risk. *Risk Analysis*, **1(1)**, pp. 11–27, 1981.

[2] Paul, B.K., *Environmental Hazards and Disasters: Contexts, Perspectives and Management*, Wiley-Blackwell: Chichester, 2011.

[3] EU, *European Union on the assessment and management of flood risks: Directive 2007/60/EC of the European Parliament and the Council*, 2007.

[4] Merz, B., Elmer, F. & Thieken, A.H., Significance of "high probability/low damage" versus "low probability/high damage" flood events. *Natural Hazards and Earth System Sciences*, **9(3)**, pp. 1033–1046, 2009.

[5] Thieken, A.H., Apel, H. & Merz, B., Assessing the probability of large-scale flood loss events: a case study for the river Rhine, Germany. *Journal of Flood Risk Management*, **8(3)**, pp. 247–262, 2015.

[6] Huttenlau, M., Stötter, J. & Stiefelmeyer, H., Risk-based damage potential and loss estimation of extreme flooding scenarios in the Austrian Federal Province of Tyrol. *Natural Hazards and Earth System Sciences*, **10(12)**, pp. 2451–2473, 2010.

[7] Ghizzoni, T., Roth, G. & Rudari, R., Multivariate skew-t approach to the design of accumulation risk scenarios for the flooding hazard. *Advances in Water Resources*, **33(10)**, pp. 1243–1255, 2010.

[8] Keef, C., Tawn, J. & Svensson, C., Spatial risk assessment for extreme river flows. *Journal of the Royal Statistical Society Series C*, **58(5)**, pp. 601–618, 2009.

[9] Schneeberger, K., Huttenlau, M., Winter, B., Steinberger, T., Achleitner, S. & Stötter, J., A probabilistic framework for risk analysis of widespread flood events: A proof-of-concept study. (in prep.).

[10] Heffernan, J.E. & Tawn, J.A., A conditional approach for multivariate extreme values (with discussion). *Journal of the Royal Statistical Society Series B*, **66(3)**, pp. 497–546, 2004.

[11] Merz, B., Kreibich, H., Schwarze, R. & Thieken, A.H., Review article "Assessment of economic flood damage". *Natural Hazards and Earth System Sciencies*, **10(8)**, pp. 1697–1724, 2010.

[12] Fuchs, S., Susceptibility versus resilience to mountain hazards in Austria - paradigms of vulnerability revisited. *Natural Hazards and Earth System Sciences*, **9(2)**, pp. 337–352, 2009.

[13] Fuchs, S., Birkmann, J. & Glade, T., Vulnerability assessment in natural hazard and risk analysis: current approaches and future challenges. *Natural Hazards*, **64(3)**, pp. 1969–75, 2012.

[14] Huttenlau, M. & Stötter, J., The structural vulnerability in the framework of natural hazard risk analyses and the exemplary application for storm loss modelling in Tyrol (Austria). *Natural Hazards*, **58(2)**, pp. 705–729, 2011.

[15] Statistik Austria, www.statistik.at/web_de/klassifikationen/regionale_gliederungen/dauersiedlungsraum.

[16] Schneeberger, K., Huttenlau, M. & Stötter, J., A seasonality analysis of flood peaks as basis for flood risk analyses in two Alpine regions. *Proc. of the 3rd STAHY International Workshop on Statistical Methods for Hydrology and Water Resources Management*: Tunis, pp. 1–11, 2012.

[17] Habersack H., Krapesch G., *Hochwasser 2005 - Ereignisdokumentation: der Bundeswasserbauverwaltung, des Forsttechnischen Dienstes für Wildbach- und Lawinenverbauung und des Hydrographischen Dienstes*, 2006.

[18] Merz, R., Blöschl, G. & Humer, G., National flood discharge mapping in Austria. *Natural Hazards*, **46(1)**, pp. 53–72, 2008.

[19] Amt der Vorarlberger Landesregierung, *HORA/Vorarlberg: Hydraulische Neuberechnung für Vorarlberg*, 2010.

[20] Huttenlau, M. & Stötter, J., Ermittlung des monetären Werteinventars als Basis von Analysen naturgefahreninduzierter Risiken in Tirol. *Geographica Helvetica*, **(2)**, pp. 85–93, 2008.

[21] Meyer, V., Schwarze, R., Becker, N., Markantonis, V., van den Bergh, J.C.J.M., Bouwer, L.M., Bubeck, P., Ciavola, P., Genovese, E., Green, C., Hallegatte, S., Kreibich, H., Lequeux, Q., Logar, I., Papyrakis, E., Pfurtscheller, C., Poussin, J., Przyluski, V., Thieken, A.H. & Viavattene, C.,

Assessing the costs of natural hazards – state of the art and knowledge gaps. *Natural Hazards and Earth System Sciences*, **13(5)**, pp. 1351–1373, 2013.

[22] Kreibich, H., Müller, M., Thieken, A.H. & Merz, B., Flood precaution of companies and their ability to cope with the flood in August 2002 in Saxony, Germany. *Water Resources Research*, **43(3)**, 2007.

[23] BUWAL, *Risikoanalyse bei gravitativen Naturgefahren: Fallbeispiele und Daten*: Bern, 1999.

[24] Oberndorfer, S., Fuchs, S., Rickenmann, D. & Andrecs, P., *Vulnerabilitätsanalyse und monetäre Schadensbewertung von Wildbachereignissen in Österreich*: Wien, 2007.

[25] Bubeck P. & de Moel, H., *Sensitivity analysis of flood damage calculations for the river Rhine*, 2010.

[26] Jongman, B., Kreibich, H., Apel, H., Barredo, J.I, Bates, P.D, Feyen, L., Gericke, A., Neal, J., Aerts, J.C.J.H. & Ward, P.J., Comparative flood damage model assessment: towards a European approach. *Natural Hazards and Earth System Sciences*, **12(12)**, pp. 3733–3752, 2012.

[27] Thieken, A.H., Olschweski, A., Kreibich, H., Kobsch, S. & Merz, B., Development and evaluation of FLEMOps – a new Flood Loss Estimation Model for the private sector. *WIT Transactions on Ecology and the Environment*, **184**, eds. D. Proverbs & C.A. Brebbia, WIT Press: Southampton, pp. 315–324, 2008.

[28] Kreibich, H., Seifert, I., Merz, B. & Thieken, A.H., Development of FLEMOcs – a new model for the estimation of flood losses in the commercial sector. *Hydrological Sciences Journal*, **55(8)**, pp. 1302–1314, 2010.

[29] Hübl, J. & Kraus, D., *Erweiterungsvorschläge zur Kosten - Nutzen - Untersuchung der Wildbach- und Lawinenverbauung*: Wien, 2004.

[30] IKSE, *Aktionsplan Hochwasserschutz Elbe*: Magdeburg, 2003.

[31] MURL, *Hochwasserschadenspotentiale am Rhein in Nordrhein-Westfalen*: Düsseldorf, 2000.

[32] IKSR, *Atlas der Überschwemmungsgefährdung und möglichen Schäden bei Extremhochwasser am Rhein*: Koblenz, 2001.

[33] Romang, H., *Wirksamkeit und Kosten von Wildbach-Schutzmassnahmen*, Geograph. Inst. der Univ. Bern, 2004.

[34] Schmidtke, R.F., Sozio-ökonomische Schäden von Hochwasserkatastrophen. *Wasserbauliche Mitteilungen der Technischen Hochschule Darmstadt* **(40)**, pp. 143–58, 1995.

[35] Romanowicz, R. & Beven, K., Estimation of flood inundation probabilities as conditioned on event inundation maps. *Water Resources Research*, **39(3)**, 2003.

Flood hazard: planning approach to risk mitigation and periphery rehabilitation

F. D. Moccia & A. Sgobbo
Department of Architecture, Naples University, Italy

Abstract

Control of risks associated with floods in recent years has a central role in scientific research on urban planning. This is related to the increased incidence of these phenomena whose root cause was found in the greater frequency of extreme rainfall events as a result of climate change. In Italy, the phenomenon became dramatic because of the numerous human casualties that occur each year. Without prejudice to the need for interventions that effectively act on climate change, there is a need to provide interventions to mitigate the consequences in the short-term. The department of Urban Planning at Naples University has conducted thorough research on the case, objectively verifying the impact that human activity has on the dramatic consequences of rainfall events, also highlighting the lack of efficiency of the traditional mitigation measures based on the construction of huge detention basins dug in the ground. This is due to three main reasons: such interventions require financial investments that local communities often do not have; the creation of detention basins cause the removal of large quantities of soil for agricultural use; the continuous increase in impervious surfaces due to urbanization makes the need to cope with the masses of water even more pressing and therefore the dry ponds already created quickly become insufficient. The present research proposes some interventions and best practices that, if adopted, can radically reduce the impact of extreme rainfall on public safety, enabling an efficient participation of private capital in investments through the use of financial leverage and fiscal incentives.
Keywords: flood risk mitigation, anthropogenic activity, impervious surfaces, urbanization, slow-release tanks.

1 Introduction

The focus in this paper is on a special type of periphery, very different from last century's urban outskirts, mainly filled with industrial plants and worker housing. It is the part of a compact city at larger distance from the retail and office cluster of the centre, housing low income people in frequently disruptive neighbourhoods lacking the public services, open spaces for exercises or entertainment.

Modern day cities have a larger expansion unifying complex metropolitan areas with many poles partly autonomous and partly connected in a whole with a radius that can reach 40 miles or more. When urbanization extends over so great a surface, it encounters a changing geography as well as developments of morphological variety.

It is here proposed that there is a need for a planning approach able to combine the geographical setting of the built environment with the urban fabric. The planning unit should be the river basin because one of the more ancient location factors is the relation of human settlement with water and because today impacts of climate change are putting more stress over sustainable storm water management requiring a new approach to the design of its infrastructure.

In Italy the phenomenon has become critical because of the numerous human casualties that occur each year due to events that also have a strong impact in terms of economic damage to the community. Studies [1] show that over the past ten years, there have been 26 floods that have resulted in deaths and destruction in Italy with 436 deaths and more than seven billion euros of direct damage. At least another eleven billion euros were lost due to the freeze on production suffered by the affected companies and because of indirect damage suffered by the companies' equipment due to the cancellation of orders for semi-finished products.

Without prejudice against the need for effective actions over the causes driving climate change, whose effects will be, however, long-term, there is a need for a strategy to mitigate the consequences of global warming.

This, in many regions, resulted in authorities responsible for management of catchment areas beginning to develop projects aimed at improving the resilience of the territory. The classical approach is, however, end-of-pipe, based on engineering techniques aimed at mitigating the effects of the phenomenon, rather than counter the reasons. The traditional tools for such an approach are the retention basins, the reconfiguration of water bodies' section and the creation of devices to contain the flow of debris.

The research at the Department of Architecture in Federico II University highlights the important role of human activity in transforming the land for their own needs arising from the dramatic consequences of rainfall events, together with the lack of efficiency of the traditional mitigation measures based on the construction of huge retention basins dug in the ground.

This approach is demonstrated in a case study with proposals for the Sarno river basin in the periphery of the Naples metropolitan area, where problems of flooding and pollution arise from the high housing density and industrial plants and great projects with the "engineering paradigm" have not yet been successful.

In fact, those who pursue a sustainable and resilient city often adopt an incremental approach: they accept as a starting point the status quo proposing changes that can improve environmental conditions; they put emphasis on actions to mitigate the situation introducing complementary elements. Pedestrian areas, green spaces in the residual areas, small plants for phyto-depuration and engineering solutions are some of their typical tools. In other words, they regard as immutable all that constitutes the current urban life, the objectives of planning or the values of the community and they try to make things a little better. But another point of view, considers this approach not effective, because the gravity of the situation requires much more radical transformations. A point of view that proposes new models for the eco-city, very different from what is commonly accepted but anyway very interesting if proposing new models is thought of as a typical urban planner task. Exploring such new models and overcoming the critics believing in catastrophism in which this position can easily result, it is discovered that the European historical cities already offer many of the needed solutions and that the required actions can be an opportunity for activities which also involve aspects of urban quality [2].

The present thesis is that planning a system of permeable green infrastructure is a tool for achieving an in-depth change in fringe metropolitan areas both in providing a number of ecosystem services and making deprived areas more resilient. The plan involves also the design of new quality urban spaces.

2 The flooding risk management issue

There are many causes combined which determine the need to cope rapidly with the risk of floods in Italian cities. First, the intense processes of urbanization of the country since the early 50s occurred without proper observance of the geographical and hydro-geological characteristics of developing sites, whose selection was made mainly because of economic considerations and the intense demand for detached houses meant that buildings were built wherever the lack of adequate local regulations allowed it.

Progressive urbanization goes hand in hand with the abandonment of the agricultural areas which remained uncultivated before being incorporated into the city. Even a high percentage of uncultivated areas have a negative impact on flood risk. In a very simple way it can be estimated that, on average, about one-third of rainfall overflows directly to the oceans; another third infiltrates into the ground either remaining in the surface layers or resurfacing in streams and rivers, or go down to deeper levels; a third, finally, returns to the atmosphere by evapotranspiration [3].

These percentages, however, are subject to many variables, including geographical location, the rainfall regime, soil type, the saturation degree of the ground, etc. For example, in a typical rural area with land used for agricultural purposes and small slope, the percentage of rainwater that infiltrates into the ground can reach 50%, 40% in evaporation while surface runoff reaches only 10%. In an uncultivated area these values are 30%, 40% and 30%, respectively, on average.

In modern urban areas the ratio of water that infiltrates into the subsoil to water flowing on the surface changes dramatically: the runoff is 55% on average but it can grow up to 80%, while the deep infiltration is reduced to values close the zero.

Thus the increase in urbanized areas and uncultivated ones results in two important consequences: it increases the amount of water that must be disposed through the streams network; it reduces soil stability and leads to a progressive deforestation.

The notable speed of water flowing into a stream can cause, when a bottleneck occurs (bridges, sudden change in direction or section), sudden and violent flooding that, most of the time, leads to deaths and damage.

Human activity on the natural environment affects the occurrence of floods for various reasons: firstly, because these actions result in climate change. The close relationship existing between human input and climate change is now broadly recognized by the scientific community. Climate change also affected the phenomena because of the significant change in the rainfall regime [4, 5].

More and more violent phenomena are witnessed, characterized by brief but very intense rainfall with large volumes of water falling on limited portions of territory.

When there is alternation between long periods of drought and times of intense rain, this has serious consequences on the drainage system.

Precipitations can be classified according to three key features: rainfall depth; duration; rainfall intensity.

Rainfall depth, h, is the thickness of the water layer which would remain in the ground if there was not any infiltration and evaporation. The unit conventionally used for hydrological measurements is the millimetre, which allows the quantification of a precipitation regardless of the surface on which it is measured.

Duration, t, is the period of rainfall inside a given rainfall intensity.

Rainfall intensity, j, is defined as the ratio of the total amount of rain falling during a given period to the duration of the period. It is expressed in depth units per unit time, usually as mm per hour, by the equation:

$$j = h/t$$

Studies on precipitation forecasts were made by analyzing the data from the rain gauges available in the basin and Capri weather station; they were the most representative of the average conditions in the Sarno river catchment area.

The results were, that in the period 1960–1980, the average rainfall daily depth was 45 mm. In 1980–2000, it reached 54 mm with a growth of 20%. In the last decade, the average rainfall daily depth was 60 mm with a growth of 33% over the 60s.

On the other hand, as highlighted in the histogram of fig. 1, the total amount of rain in a year has actually decreased, confirming the general trend of less rain in a few intense storms.

Urbanization itself has a considerable effect on the rainfall regime resulting in more intense events as shown by researchers at the University of Arkansas and NASA's Goddard Space Flight Center [6] in a study to quantify the impact of urbanization on meteorology in the areas surrounding the city of Houston, for which the actual existence of a close correlation between urban density and intensity and frequency of rains was verified.

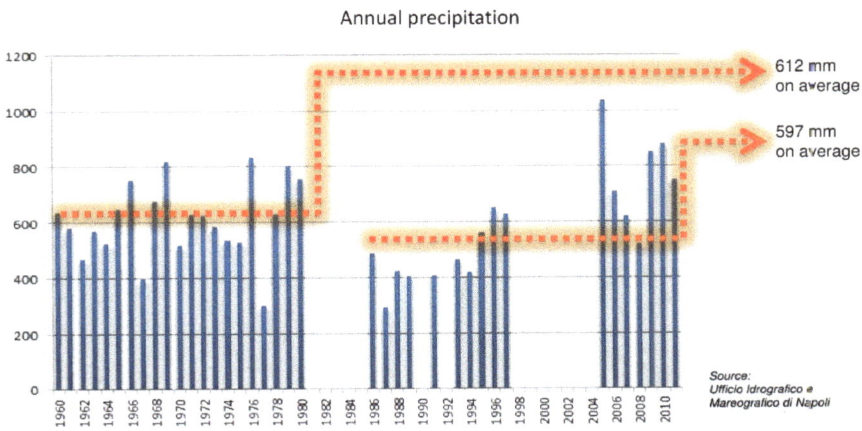

Figure 1: Rain annual performance.

3 Land use and urbanization process

The territory relating to the former Sarno Basin Authority (fig. 2) is bordered to the north, by the slopes of Vesuvius, to the South, by the Sorrento Peninsula, to the west by the Gulf of Naples and to the East, by the mountains of Solofra and Pizzo of Alvano.

It extends over an area of approximately 714 square kilometers and includes 60 municipalities. The analysis of population dynamics in the area between 1971 and 2011, as well as the population density showed a general increase in the number of inhabitants from a population of 1,102,242 to the current 1,163,715. The population density is consistently very high, from 1,543 inhabitants/sq km in 1971 to 1,629 inhabitants/sq km in 2011.

The foothill areas in the territory are heavily transformed for the human uses and activities and, often, the construction of settlements, more or less authorized, has also affected areas directly facing the streams and drainage canals of the basin; the result is a situation of risk for flooding that has become intolerable [7].

The mountain streams are normally completely dry, being only rarely affected by the presence of small natural springs. The presence of water is, therefore, observable only in rainy weather. As a result of the presence of debris and the occasional runoff of the slopes, the sediment supply to these streams is quite high. In the absence of maintenance, this results in a state of danger related to the possible formation of rivers of mud.

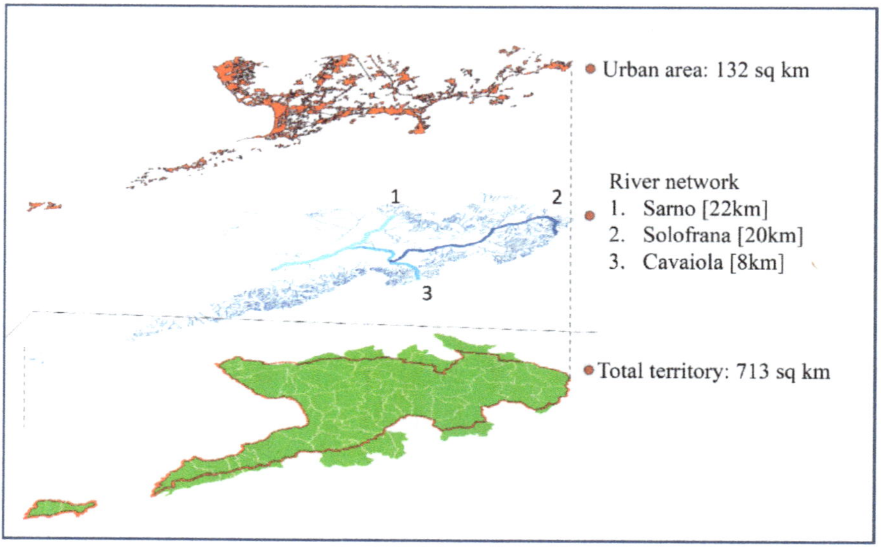

Figure 2: The Sarno basin.

Urbanization in the Sarno river catchment area was disruptive and this went together with progressive abandonment of the countryside.

Data provided by local associations of farmers showed that until 1980 the area used for agricultural purposes inside the Sarno river catchment area stood at 119,448 acres. In 2010 this value was reduced to 48,034 acres, with a 60% loss.

What, however, is more surprising is that cultivated countryside abandonment is accelerating. In 1980-1990 the loss was 15.3%. In the following decade, 16.9% of what remained has been lost. In 2000-2010, the reduction of area designated for agricultural use was 42% of the 84,120 acres that remained at the start of the period.

Finally, the urban consumption of agricultural land reaches an overall amount which corresponds to the entire surface of 24 municipalities inside the basin.

For rivers, embedded in the surrounding alluvial plain, risk situations are those that arise primarily from the presence of bottlenecks, such as the numerous bridges often poorly designed, from poor maintenance of the banks and flood walls and because of numerous parts artificially covered in which occlusions and phenomena of pressure peak may occur.

4 Causal links between flood risk and imperviousness

The progressive reduction of soil drainage capacity, due to the construction of roads, buildings and, in general, to the process of imperviousness, leads to a considerable increase of the runoff coefficient [8].

The runoff coefficient Φ is the ratio between the volume of water that reaches the drainage system (w_{inf}) and the total volume of rain (w_{tot}) [9] according to the following equation:

$$\Phi = \frac{W_{inf}}{W_{tot}} \tag{1}$$

In the Sarno river basin, the total area today urbanized represents approximately 18% of the entire basin and the runoff coefficient, in the period 1970–2010, increased from 0.21 to 0.43.

The calculation of the runoff coefficient was achieved primarily by building a model of soil that was representative of the actual conditions encountered in the area. For this purpose, the whole of the basin has been divided into homogeneous areas in relation to which a virtual runoff coefficient was calculated, capable to represent numerically the contribution per unit of each type of urban fabric and rural area in the basin.

In the case analyzed by the research, the following site types were identified: historical city center, sprawl, isolated building area, compact urban precinct, agricultural area; forest; industrial and commercial district; large specialized settlement.

The size, in square kilometres, of every area existing in the basin was calculated and each homogeneous area has been studied by determining the weighted average extent of every elementary surface components.

The elementary surface components are classified into: impervious roads, half impervious paving roads, impervious outdoor public spaces, half impervious outdoor public spaces, gardens, impervious outdoor private spaces, flat impervious roofs, impervious sloping roofs, green roofs, impervious forecourts, half impervious forecourts, (indicated as grey fields in the Congress for the New Urbanism [10]), green fields, uncultivated and abandoned fields. The whole basin, finally, has been divided into three main complex areas in relation to the substantial difference of the conditions in urbanization and gravitation to the main river: a coastal area, composed by the municipalities directly overlooking the sea that, although highly urbanized, acts at particularly low flow rates on the Sarno disposing directly into the sea most of the surface water; an internal urban area, located close to the three main channels of the basin; a hilly rural area, consisting of small towns, predominantly hilly and sparsely urbanized.

Focusing the analysis on the internal urban area and studying its elementary surfaces, it has been possible to identify the characteristic composition of each type of area (see table 1 for an example of historical city center areas analysis).

Having obtained the runoff coefficient relating to the various surface types from the literature (DIN standards, Frühling scale and UNI 11235:2007), the runoff coefficient of the model areas was calculated as the weighted average of the component surfaces. Finally, for each surface, the percentage of contribution to the basin was also determined; this represents the part of the water collected that is actually disposed by the river and streams network.

The traditional approach to urban drainage involves the collection of all the rainwater falling on the surfaces, which are more or less impervious. This water is placed in a disposal system which, in the best of conditions, provides a drain for the ordinary waste water, one for gray water and one for the sewage. Very often the drains for gray water and sewerage are reduced to a single system connected to a sewage treatment plant while rain water is discharged directly into drainage system or sea. In the regions of southern Italy, the focus of the Department of Urban Planning research, the sewer system is actually almost always a combined waste water system. In fact, each of the above systems presents some fragility.

Table 1: The urban fabric historical city center.

Historical city center	Torre Annunziata	Pompei	Sarno	Typical area
				Average
	%	%	%	%
Impervious roads	6.95%	4.80%	9.15%	**6.51%**
Half impervious paving roads	4.00%	7.65%	3.85%	**5.83%**
Impervious outdoor public spaces	6.05%	3.40%	4.20%	**4.46%**
Half impervious outdoor public spaces	4.85%	8.25%	3.85%	**5.88%**
Gardens	5.20%	6.15%	4.10%	**4.68%**
Impervious outdoor private spaces	14.50%	14.95%	18.50%	**16.54%**
Flat impervious roofs	36.25%	27.75%	25.80%	**27.99%**
Impervious sloping roofs	9.85%	11.60%	16.50%	**14.14%**
Green roof	0.00%	0.00%	0.00%	**0.00%**
Impervious forecourt	3.60%	3.45%	5.70%	**4.06%**
Half impervious forecourt	0.85%	2.60%	0.55%	**1.45%**
Green fields	3.55%	6.00%	3.60%	**3.98%**
Uncultivated and abandoned fields	4.35%	3.40%	4.20%	**4.50%**
	100.00%	100.00%	100.00%	**100.00%**

Separate sewers and double sewer systems have the advantage of reducing considerably the amount of material to be treated in sewage treatment plants. However, the mistaken belief that the water collected from ordinary urban surfaces can be considered free of pollutants causes quite a few negative effects on the coastal and river aquatic environment.

The combined sewage has the advantage of ensuring a treatment for all types of water but, on the occasion of heavy rainfall, when water masses that reach the treatment plant are far higher than the actual plant capacity, part of the sewage is directly poured into the sea as is [1].

Every urban fabric type has also been studied assuming that all the rainwater would be disposed by the river, as suggested by the draft from the Sarno Basin Authority. In this case, (see table 2), for example, the urban fabric categorised as historical city center, results in a runoff coefficient enormously increased, from a value of 0.43 to an amazing 0.75 on average.

One of the main negative effects of the urbanization process on the floods problem may be detected from the short time in which a drop of rain reaches the disposal system. This time, normally identified by calculating the time of concentration, varies depending of topographic and geological characteristics of the basin, but, above all, of the roughness and permeability of surfaces [11].

Table 2: Runoff after disposal network revision.

Historical city center	Relative weight	Surface runoff coefficient	Contribution to the basin	Virtual runoff coefficient
	%		%	
Impervious roads	6.51%	0.90	100.00%	0.06
Half impervious paving roads	5.83%	0.70	100.00%	0.04
Impervious outdoor public spaces	4.46%	0.90	100.00%	0.04
Half impervious outdoor public spaces	5.88%	0.70	100.00%	0.04
Gardens	4.68%	0.15	100.00%	0.01
Impervious outdoor private spaces	16.54%	0.85	100.00%	0.14
Flat impervious roofs	27.99%	0.85	100.00%	0.24
Impervious sloping roofs	14.14%	0.95	100.00%	0.13
Green roof	0.00%	0.70	100.00%	-
Impervious forecourt	4.06%	0.85	100.00%	0.03
Half impervious forecourt	1.45%	0.60	100.00%	0.01
Green fields	3.98%	0.10	100.00%	0.00
Uncultivated and abandoned fields	4.50%	0.25	100.00%	0.01
		Area virtual runoff coefficient		**0.76**

The concentration time, evaluated at a given point of a network of drainage (natural or artificial), is the elapsed time for the generic raindrop to reach the closing section of the basin, starting from the farthest point from the hydraulic point of view. It represents a useful measure for predicting the flow of the watercourse that drains a basin resulting from hypothetical rainfall events.

The fact that the surface roughness is significant in determining the time of concentration leads to a further important observation: the historic city was not only smaller in size but, with regard to flood risk, had an intrinsic greater resilience [12].

In the study of the time of concentration, it is very important to observe the characteristics of rainfall intensity and its evolution over time. As already mentioned, probably because of climate change, rains characterized by very high intensity in short periods are being seen in our latitudes.

In the case of the Sarno river catchment area the analysis of available rainfall data has shown that, on a return period of 100 years, the significant lifetime of storm is about 21 minutes. This means that, if it were possible, for a given area, to increase the time of concentration measured at the beginning of urban disposal network to 21 minutes, this surface practically would not participate to the flood risk.

5 River basin authorities action plan and green infrastructure alternative

The traditional flood risk mitigation system is based on end-of-pipe tools and the only prevention activities carried out on the drainage system are limited to over-dimensioning pipelines and constructing river banks. The most widely used flood retention system, downstream of the drainage system, is the construction of huge detention basins dug in the ground. As noted by the relevant scientific literature, however, these measures often fail to cope effectively with the amount of water resulting from more intense rainstorms demonstrating their usefulness only in those cases which, by themselves would have caused little damage [12, 13].

The system aims to offer a space to accommodate temporarily the masses of excess water in order to allow the rivers and streams to dispose of an amount compatible with their capabilities.

The river basin authorities, after a long period of evaluations, have recently approved an action plan that mitigates recurrent flooding with 9 new retention basins. The overall cost, estimated at about 250 million euros, is financed almost entirely by the European Regional Development Fund (ERDF 2007–2013).

The main weaknesses of the project have been identified as: exclusively public investment, which does not involve citizens and felt by people as an imposition; impact on agricultural land that implies about 196 acres lost. In a valley already crowded by a dense population some of whom is still devoted to intensive and high quality agriculture, the loss of such an amount of productive land raised some criticism and a growing local opposition.

The research conducted by the Naples University highlights the lack of efficiency obtainable from this kind of system. This is due to three main reasons: such interventions require huge financial investments that local communities often do not have; the creation of dry ponds cause, as a result, the removal of large quantities of soil for agricultural use; the steady growth of impervious surfaces due to the phenomena of urbanization makes the need to cope with the masses of water ever more pressing and therefore the detention basins already created quickly become insufficient.

The complexity of the city implies that a correct approach aimed at reducing the impact that urban areas have on the natural water regime and, in particular, on the management of surface water, can only be a multiplicity of interventions at different scales [14]. There is also the need to consider that the actors and stakeholders acting in the city system are numerous and motivated by different purposes.

Another significant element is that, by now, the phenomena of urbanization and urban sprawl have reached levels far exceeding the tolerance threshold [15] and, therefore, the mitigation measures to be taken cannot be absolutely limited to a virtuous design of expansion areas but they must necessarily also affect the already compact urban fabric.

Therefore, the toolset available to urban planner has recently been enriched by several new best practices. Those aimed at new development, in general, include infrastructure elements such as street side swales (trenchless drain), the use of paving that can ensure good permeability, the waste water disposal systems interwoven with artificial lakes that perform the dual function of landscaping and of regulation elements of water masses to be disposed. At the same time, they include rules of construction for private developers such as the prohibition of certain types of buildings, the obligation to implement green roofs or to use specific precautions in the management of land devoid of buildings [16].

Measures aimed at the fabric inside urban precinct are in turn grouped into two subgroups: public initiative interventions and private initiative interventions. The first group, although rich in many proposals and extensive studies, is characterized by a very low number of direct experiences because of the particularly high costs borne by state finances and the difficult relationship between public and private property which is often an insurmountable obstacle, especially in countries where the legal tradition is based on Roman law.

The second group is, mostly, micro interventions at the scale of single ownership or single building, in which the role of the planner becomes central in designing a proper relationship of balance between incentives and effort required. Often, this balance is influenced by ideological conflicts that may cause either excessive burdens for the private owner (the consequence of which is the lack of action) or a set of rules too gentle (which leads to low actions efficiency compared to target).

The authors' department research on the Sarno river basin suggests some solutions whose effects, assessed via a model, are differentiated according to the various surfaces of every analyzed type areas.

For impervious roads, outside urban precinct, trenchless drain seems to be the best solution. Side swales are designed so that the amount of water that can be stored for long term infiltration is about that corresponding to what can flow over the street in 21 minutes. Width and thickness of trenchless drain is designed by using the following equation:

$$W = \frac{k_1 jL}{3k_2 kT^n - k_1 j} \qquad (2)$$

where W is the trenchless drain width, L the road width, k the hydraulic conductivity of soil, k_1 and k_2 are coefficients representing losses, generally <1, n a time exponent depending on local conditions, T the time in which water remains inside the swale and j the rainfall intensity.

This means that, if, on average, roads inside the basin are 12 m wide and rainfall depth, in 21 minutes, is about 71 mm, trenchless drain has to store water for 0.85 mc for every one-meter length. To reach this amount, it is possible to build side swales 2 meters long that, under a thin ground layer, are composed with a gravel layer of about 50 cm thickness.

Inside the urban precinct, side swales are difficult to implement. In this case, where it is not possible to build avenues with large permeable areas such as in the Oregon's Green Streets Program [17], a solution using an absorbing drainage network is progressively implementable. The first step involves changing the street manholes to bottomless absorbing manholes. In this case, however, the capacity to reduce the inflow is strongly influenced by the geological characteristics of the soil and, above all, from its intrinsic porosity. In the long term, it is possible to progressively change impervious paving with pervious ones, especially in the side street parking. This, in fact, not only acts on the absorbing capability but, above all, slows down the time of concentration.

The strategy for larger impervious outdoor public spaces, such as squares and public facilities, mainly concern interventions on the pavement, by progressively adopting solutions with high permeability and adequate integration of green absorbent spaces together with dispersants wells.

In the tested solution, absorbing wells are provided; one for every 1,000 square meters of area, aimed at the disposal of rainwater and located downstream the compulsory treatment of the first 5 mm water flowing over the surface, aimed at reducing the charge of hydrocarbons and heavy metals. The process of filtration through the soil will ensure, in any case, the further purification of the fluid from any residual contaminants before it reaches the groundwater.

Of course, the infiltrating wells are necessarily designed in relation to the specific characteristics of the area in which this is realized, however, changing even within the same city. The first 5 mm of rainwater, after purification, is still entrusted to the urban drainage system. The wells, slow-release, are dimensioned with a minimum volume of about 10 cubic meters, are deepened until reaching a level consisting of layers sufficiently permeable and have a diameter of between 20 and 30 cm. They have one overflow drain able to dispose of excess water during storms of exceptional intensity.

For built areas, the proposed solution is the incentive for the construction of green roofs. These, in fact, have the advantage of ensuring an efficient filtering action of rainwater that significantly reduces the level of pollutants and dangerous waste by avoiding the need for further downstream processing. At the same time, they significantly increase the time of concentration by reducing the instantaneous quantity of water transmitted to the urban system of disposal. The best results are obtained implementing also small tanks at the foot of the individual downspout. In general, the proper technique for the design of rain water drains provides that each vertical pipe serves an area of about 100 square meters. The available studies on precipitations in southern Italy showed that, to prevent water falling on the roofs reaching the urban disposal system before the intensity of the storm has been reduced, it is enough to ensure a collection capacity of about 3.85 cubic meters.

The product designed as part of the research program is constituted of a tank of recycled plastic with two exhaust systems. One, always active, has a flow rate of 2.4 l/min and allows the complete emptying of the tank in about 26 hours and 45 min. The other, with functions of overflow, having a flow rate of 160 l/min., allows the disposal of the excess water to the white water collection system.

In summary, the experimental project pursued during the academic research for the mitigation of flood risk proposes: creation of 490 acres of green roofs; management through the slow-release tanks of water coming from 4,450 acres of roof; creation of 57 km of street side swales for disposal of the water coming from roads; 4,080 acres of impervious surfaces managed through systems of absorbing wells (fig. 3). This results in decreasing the urban areas basin runoff coefficient to 0.32 and the overall runoff decreases by about 1.5 million m^3/year.

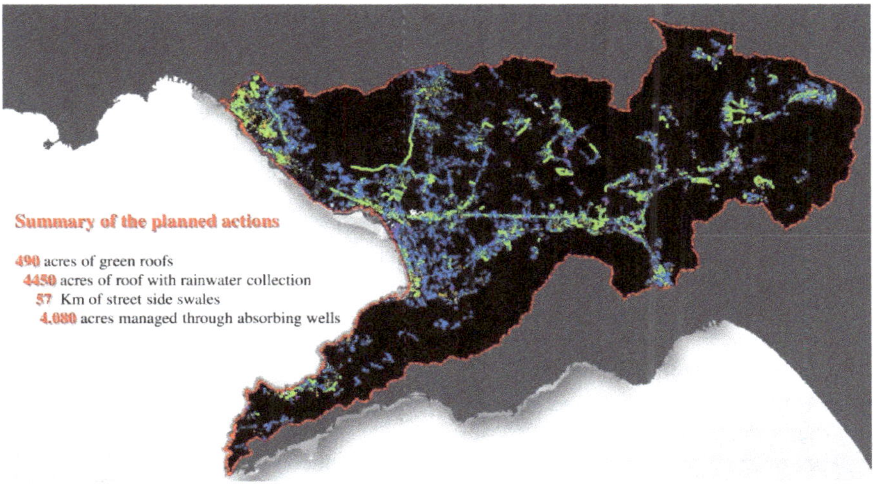

Figure 1: Localization of designed solutions.

Academic research has also investigated the viability of the project. The overall work costs about 460 million euros. However, only 95 million euros come from public budget. The remaining part is borne by private sources and encouraged through fiscal measures and incentives to implement building transformations.

Fiscal incentives are incorporated into an already existing law that provides for buildings maintenance efficiency and sustainability improving the possibility to deduct 55% of costs from taxes. Therefore, the part of private investment that can take advantage of deductions is about 271 million euros, resulting in leverage fund of about 150 million distributed, however, in the following ten years.

The incentives to operate building transformations concern the possibility of expansion and integration of different uses. This also allows further investments with predictable advantages on the overall city sustainability.

6 Conclusions and side effects

One of the most interesting aspects in the solutions proposed and tested in the authors' departmental research is linked to the substantial positive side effects registered on the urban quality in the degraded southern Italy suburbs.

The first results, in fact, seem to reinforce the idea that an urban planner approach to the issue of urban resilience offers a good opportunity for high quality city regeneration.

In brief, among the many positive side effects achieved, soil formation is pointed out: green roofs give the opportunity to recover a part of lost agricultural land thanks to the implemented solutions effectively providing little urban farms; climate regulation and prevention of urban heat island: green roof, side swales and green parking lots offer a substantial increase of green surfaces and, together with increasing of pervious paving, the opportunity of a natural regulation of urban heat in the summer thanks to increased shading and evapotranspiration; aesthetic and quality urban spaces: green roofs and increased green public areas provide the opportunity to mitigate the "grey effect" typically generated in the south of Italy by peripheral urban landscape and recreate the lost sense of place; citizen's sense of belonging: direct participation in investment and transformation improves the sense of belonging and the desire to protect the regenerated urban space.

Urbanity, in its physical structure and morphology, depends on the design of infrastructure, because infrastructures are a service but also an urban space. It means that working on the infrastructural network works also on the quality of urban space and this can be achieved at different levels.

The minimal effect is achieved in the maintenance of urban space: when new green parking lots are built, deteriorated space is also being regenerated; when a new permeable paving is built, road maintenance is also carried out. In fact, a green infrastructure network is economically viable only if it is part of a general action of urban maintenance.

However, when green infrastructure is introduced into urban fabric, new services to the community are being also created and lack of services is a strong characteristic of today peripheral metropolitan areas. The latter were, in fact, developed with an approach based on the landowners' and developers' interest, often without urban planning with cheap houses jerry-built, regardless of services and public spaces.

Investing funds intended for construction of retention basins for the creation of an intra-urban green infrastructure network allow, therefore, the multiplication of the positive effects obtained with such investment by regenerating deprived urban areas.

In this way, transforming the resilience goal from cost into opportunities becomes the new strategy to address the issue of metropolitan suburbs meaning with it not only to fill gaps but to conceive a new idea of city: low impact, healthy, eco-friendly, able to meet the modern civic and social inhabitants' sensibilities. Such a strategy, in fact, is also able to find favor with urban social movements first of all involved in the public health problems [18].

This even means the reversal of the subordination of fringe metropolitan areas to the center. In fact, the greater amount of natural spaces and less congestion of existing suburbs allow the achievement of a healthy city model far more easily than downtown.

Finally, research has shown that the conscious use of already existing fiscal incentives can generate a virtuous leverage on public investment required to address a problem of such a large scale as the suburbs regeneration.

The incentives to implement building transformations concern the possibility of expansion and integration of different uses. This also allows the promotion of further investments with predictable advantages in terms of overall efficiency and sustainability of the city.

References

[1] Moccia, F.D. & Sgobbo, A., Flood hazard: planning approach to risk mitigation. *Safety and Security Engineering V,* eds. M. Guarascio, C.A. Brebbia & F. Garzia, WIT Press: Southampton, pp. 89–99, 2013.

[2] Moccia, F.D., City planning in climate change times. *Urbanistica,* **LXI (140)**, pp. 95–99, 2009.

[3] Montin, P., *Acque meteoriche di dilavamento. Principi di progettazione e dimensionamento degli impianti di trattamento,* Dario Floccavio Editore: Palermo, Italy.

[4] Bernhofer, C., Franke, J., Goldberg, V., Seegert, J. & Kuchler, W., Regional climate change. To be included in future flood risk analysis? (Chapter 8). *Flood Risk Management: Hazards, Vulnerability and Mitigation Measures,* eds. J. Schanze, E. Zeman & J. Marsalek, Springer: Dordrecht, The Netherlands, pp. 93–96, 2006.

[5] Polemio, M. & Lonigro, T., Variabilità climatica e ricorrenza delle calamità idrogeologiche in Puglia. *Le modificazioni climatiche e i rischi naturali. Atti delle giornate studio,* ed. M. Polemio, CNR IRPI: Bari, Italy, pp. 13–15, 2011.

[6] Shepherd, J.M. & Burian, S., Detection of urban-induced rainfall anomalies in a major coastal city. *Earth Interaction,* **7**, pp. 1–17, 2003.

[7] Braioni, M.G., Villani, M.C., Braioni, A., Salmoiraghi, G., Locascio, A. & Cannata, G., The restoration of some stretches of the Sarno River (Southern Italy): from canalized environment to fluvial corridor. *River Easin Management V* , ed. C.A. Brebbia, WIT Press: Southampton, pp. 293–304, 2009.

[8] Rimoldi, P., *Ingegneria idraulica urbana. Acquedotti e fognature. Manuale-tecnico pratico per la progettazione e la realizzazione delle opere,* Maggioli editore: Santarcangelo di Romagna, Italy, p. 172, 2010.

[9] Mishra, S.K. & Singh, V.P., *Soil Conservation Service Curve Number (SCS-CN) Methodology,* Kluwer Academic Publishers: Dordrecht, The Netherlands, pp. 61–63, 2003.

[10] Global Strategic Real Estate Research Group, Greyfield Regional Mall Study. In Congress for the New Urbanism, PWC. Retrieved from www.cnu.org.

[11] Punmia, B.C., Ashok, K.J. & Arun, K.J., *Waste Water Engineering,* Firewall Media: New Delhi, India, 1998.

[12] Schumann, A.H., *Flood Risk Assessment and Management: How to specify Hydrological Loads, their Consequences and Uncertainties,* Springer: London, U.K., pp. 260–263, 2011.

[13] Thorsteinsson, D., Semadeni-Davies, A. & Larsson, R., Planning for river induced floods in urban areas (Chapter 25). *Flood Risk Management in Europe: Innovation in Policy and Practice,* ed. S. Begum, M.J.F. Stive & J.W. Hall, Springer: London, U.K., pp. 492–499, 2007.

[14] Martincigh, L., *Strumenti di intervento per la riqualificazione urbana. La complessità dell'ambiente stradale*, Gangemi Editore: Rome, Italy, pp. 24–26, 2012.
[15] Ercolini, M., *Cultura dell'acqua e progettazione paesistica*, Gangemi Editore: Rome, Italy, pp. 91–92, 2010.
[16] Wagner, I., Marsalek, J. & Breil, P., *Aquatic Habitats in Sustainable Urban Water Management: Science, Policy and Practice*, Routledge: London and New York, pp. 49–56, 2008.
[17] Lukes, R. & Kloss, C., *Green Streets in Managing Wet Weather with Green Infrastructure*. Municipal Handbook. Retrieved from http://water.epa.gov/infrastructure/greeninfrastructure/upload/gi_munichandbook_green_streets.pdf.
[18] Paba, G., *Movimenti urbani. Pratiche di costruzione sociale della città*, Franco Angeli: Milano, Italy, 2003.

Flood risk assessment and management: a case study in Rio de Janeiro

M. G. Miguez[1], A. P. Veról[1] & L. Bianchini[2]
[1]*Universidade Federal do Rio de Janeiro, Brazil*
[2]*Universitat Politecnica de Catalunya, Spain*

Abstract

Developing countries suffer from the growth of the so-called irregular cities. The lack of basic infrastructure and the inadequate urban growth management lead to urban flood problems. The concept of risk combines the probability of a hazard event with the consequences inflicted to a system. Thus, there are several possibilities of acting in flood risk mitigation: reducing the hazard and/or the system vulnerability. In this context, both structural measures and non-structural measures may be used. Traditional structural measures were frequently associated with river canalisation, dykes and dams. In the last decades, the concepts of sustainable drainage systems have arisen, tending to control on-source runoff generation. Storage and infiltration measures are planned to diminish the impact of the urbanisation process on the hydrological cycle. Flood mapping, land use regulation and flood proofing constructions appear in this context as non-structural measures. More recently, the concept of river restoration appeared as an alternative. Rivers in more natural conditions and associated with a responsible land use tend to produce an environment with minor risks. In this context, this paper will discuss flood risks related to an urban environment and the choices related with the attempts to mitigate the problem. The Flood Risk Index (FRI), developed by Zonensein *et al.* (11th Int. Conf. on Urban Drainage, 2008) is used to support a quantitative comparison of different design solutions. A case study concerning Mesquita City, in the metropolitan region of Rio de Janeiro, Brazil, is carried out. The urban occupation of the city is very dense, with few open spaces and degraded areas. In the overall assessment, river restoration concepts, although limited by the built environment, produced a minor flood risk index, when compared with a traditional alternative.
Keywords: flood risk, risk assessment, river restoration.

1 Introduction

Developing countries such as Brazil suffer from the growth of the so-called irregular cities comprising slum areas. The lack of basic infrastructure and the inadequate urban growth management related to these areas is generally critical and one of the main consequences is that related to urban floods. As could be observed with past experiences, traditional engineering solutions alone are not able to solve this problem. Besides, historical urban development experiences have not always looked to drainage problems. Although the urbanisation process usually aggravated floods [1], it was not common to see the integration of new developments with sustainable urban drainage concepts. However, integrated actions over the urbanised watershed, aiming the urban space revitalisation and, when possible, recovering the associated natural environment and restoring river functions, could be achieved by incorporating hydraulic functions to urban elements [2].

Hydraulic engineering works usually consider an intrinsic safety factor associated with the return period of a design event. This is a common practice that equates a certain hydraulic need regarding the considered recurrence. However, this consideration is related with the probability of occurrence of a certain hazard event, which is certainly associated with the risk involved, but is not the risk itself. The concept of risk [3] combines the probability of a hazard event with the damage that is inflicted to the system subjected to this hazard. In this context, this paper will discuss flood risks related to an urban environment and the consequences of the choices related with the attempts to mitigate the problem.

Flood risk management is an approach for dealing with flood risk based on the notion that risks cannot be entirely taken away, but only partially reduced and always at the expense of other societal goals [4]. The aim of flood risk management is thus to reduce the consequences of floods, in ways that balance this aim against other considerations. The main purpose of this paper is to demonstrate, through a specific case study, that flood risks may be effectively reduced when considering river restoration aspects. The Flood Risk Index (FRI) developed by Zonensein *et al.* [5] is used as a support tool for quantitative comparison of scenarios design solutions. The interest area in this case refers to Mesquita City, in the metropolitan region of Rio de Janeiro. The urban occupation of the city is very dense, with few open spaces. This characteristic asks for a solution that is concerned about social urban space use, valorising degraded areas and giving it different possibilities of use.

2 Risk assessment methodology

Traditionally, flood risk assessments face some difficulties due to the subjective nature of the evaluation, to the large number of factors that interfere with risk and to the non-existence of a standardised methodology. Flood Risk Index (FRI), a quantitative multi-criteria index, ranging from 0 to 100, is able to conjugate sub-indices related both to flood features and to local vulnerability and exposure characteristics. The application of the FRI requires the discretisation of the region of interest, i.e. an urban catchment, into small areas where the value of each indicator can be considered homogeneous.

The selected indicators to reflect flood properties include flood depth, a velocity factor (resulting from the maximum product of the water depth by the flow velocity), and a duration factor, associated with the permanence of the water level above a certain stage. In addition, indicators of dwelling density, income per capita, traffic and inadequate sanitation represent the estimated consequences of the flood, both tangible and intangible. Each of these factors must be converted into a scale ranging from 0 to 100, according to specific normalisation functions. The relative importance of the indicators and sub-indices is expressed by weights associated to each of them. Finally, the formulation of FRI is composed by weighted summations and products, as presented in Zonensein *et al.* [5]. FRI adapted to this work was formulated as

$$\mathrm{FRI}=\underbrace{\left[I_D \cdot p_D + I_{DF} \cdot p_{DF}\right]^{q_{FP}}}_{\mathrm{FP}} \cdot \underbrace{\left[I_{DD} \cdot p_{DD} + I_{In} \cdot p_{In} + I_{IS} \cdot p_{IS}\right]^{q_C}}_{\mathrm{C}} \qquad (1)$$

where FP is a flood properties sub-index (ranges from 0 to 100); C is a consequences sub-index (ranges from 0 to 100); I_i: indicator i (ranges from 0 to 100); p_i: weight of indicator i (ranges from 0 to 1); q_i: weight of sub-index i (ranges from 0 to 1); the meanings of the subscripts are: *D*-depth; *DF*-duration factor; *DD*-dwelling density; *In*-income per capita; *IS*-inadequate sanitation. In this specific case, the indicators velocity factor and traffic were not considered in the formulation of FRI.

It can be observed that if either of the sub-indices is null, the FRI equals zero. It should be highlighted that the choice of indicators is not necessarily unyielding and may vary according to local features or to data availability/precision. On the other hand, the proposed formulation is applicable regardless of such particularities. This index constitutes a decision support tool that allows the determination and comparison of critical zones and the assessment of the efficiency of different flood control measures, among other potential uses.

3 Flood risk mitigation

There are several possibilities of acting in flood risk mitigation. It is possible to work on the reduction of the hazard, what tends to be the most common action. Another possibility, however, is to diminish the system vulnerability. Both paths may be effective and the combination of both types of actions is desired, in order to have better and lasting results.

Two types of measures are considered when discussing flood risk mitigation: structural measures, usually related to controlling the hazard, which means reducing flood events critical characteristics; and non-structural measures, usually related to reducing exposure and vulnerability or increasing resilience. Traditional structural measures frequently were associated with river canalisation, dykes and dams. In the last decades, the concepts of sustainable drainage systems have arisen

[6, 7], tending to consider distributed measures over the basin, controlling on-source runoff generation. These types of measures are planned to diminish the impact of the urbanisation process on the hydrological cycle [8]. Storage and infiltration measures take an important place in these arrangements. The concept of a sustainable drainage system also combines non-structural measures in the proposed design solutions. Some examples of this type of actions refer to flood mapping, land use regulation and flood proofing constructions.

More recently, another concept is also being considered in the flood risk scenario: the river restoration alternative [9–11]). This alternative usually considers four pillars, with minor differences among the distinct approaches, which stand for maintenance of the natural hydrologic cycle and hydraulic risk control, fluvial morphology equilibrium, water quality and fluvial ecosystems restoration. Rivers in more natural conditions, when associated with a responsible land use planning, tend to produce a more harmonious environment, with minor risks. However, when talking about an urban area, it is very difficult to apply river restoration concepts. There are several limitations associated with the heavy watershed changes produced by the city growth. Nevertheless, it is possible to introduce some of the basic concepts, in order to have some environmental positive results, while controlling floods and diminishing city vulnerability. Recovering the connection of the river with its flood plains, relocating people in risky conditions, especially the ones living in sub-habitations over the river banks, acting over the watershed with sustainable urban drainage measures to control runoff generation (in order to provide an imitation of the natural water cycle results), may be some important actions to be held in the urban context. These measures may represent just a minor step towards an effective river restoration, but they may be a gain in the built environment.

4 Case study

Baixada Fluminense is an important region of the metropolitan area of Rio de Janeiro, Brazil. This area is known by its lowlands that spread over Iguaçu-Sarapuí River plains. Dona Eugênia River catchment (fig. 1) has a drainage area of 18 km^2 and is 10 km long, running down the hills until it reaches Sarapuí River, at the coastal plains. The first 4 km are in Nova Iguaçu City. The river spring is in Nova Iguaçu Natural Municipal Park, at an environmental preservation area called Gericinó. The subsequent 6 km are inside Mesquita City urban tissue. Throughout its 10 km length, the river passes through differentiated reaches, in which nature still presents itself preserved, in its upper reaches, and also presents degraded reaches, in the urban part of the catchment.

The FRI was applied in the case study of Dona Eugênia river basin, using census data from the Brazilian Institute of Geography and Statistics [13]. The other data source consisted of flood simulation results carried out with the aid of MODCEL, a hydrodynamic and distributed hydrologic model, suitable for urban floods simulation [14, 15]), regarding a 25-year return period rainfall event.

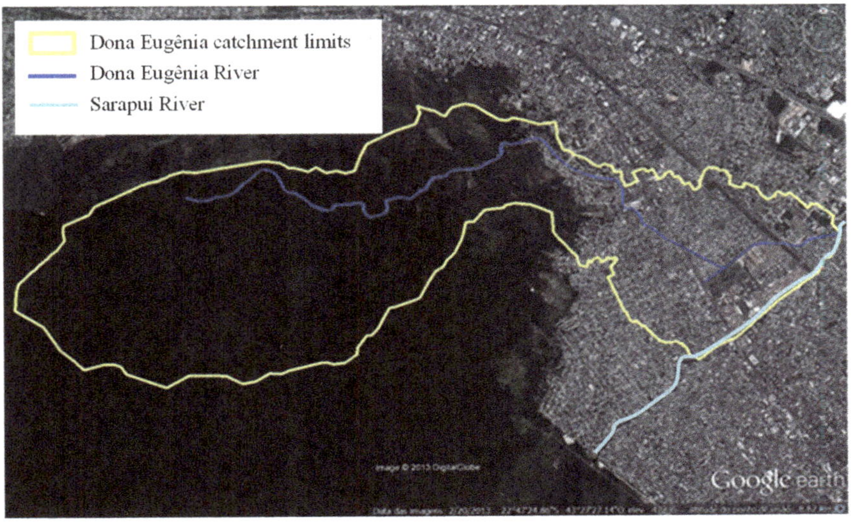

Figure 1: Dona Eugênia River Catchment [12].

Weights were determined, in a first approach, using the experience of the authors, simulating the preferences of a decision maker. In general, there are three limiting factors for the applicability of the FRI: availability of data, existence of data with adequate precision and limits of the normalisation scales. If data with adequate precision is available, the FRI is mostly limited by the normalisation of its indicators. As presented in Zonensein [5], the FRI is restricted to areas with conditions of dwelling density and income per capita similar to those found in the city of Rio de Janeiro, which was the basis for the construction of normalisation scales for these indicators. When applied to areas with significantly different conditions, which is the case of this study, the choice of indicators and their normalisation scales should be reviewed. However, the index structure is still valid, with no restrictions, regardless of the domain. Considering this, FRI was adapted to this case study. Table 1 presents the normalisation for the *Depth* indicator, associating its values with a scale of importance varying from 0 through 100. Between the established thresholds, linear interpolation may be applied.

To calculate the *Duration Factor*, it was considered that t_{10}, t_{30} and t_{50} are the times (in minutes) during which an area remains flooded with over 10, 30 and 50 cm, respectively, and that these intervals (and associated depths) are representative of the long-term effects of a flood over pedestrians, vehicles and residences, in that order.

The *Dwelling Density* indicator intends to be an estimate for the amount of people and assets affected by flood. It was considered that 0 dwelling/km² is the value with the least potential to suffer damage during floods and that 5,000 dwellings/km², which corresponds to the threshold of the variable's third quartile in the city of Mesquita, is associated with the maximum value. A linear interpolation was applied between these extremes. The *Income per capita*

Table 1: Flood depth normalisation thresholds (adapted from Zonensein *et al.* [5]).

Depth (cm)		Effect
<10	0	The curb is usually from 10 to 15 cm height, thus 10 cm of water is limited to the streets.
25	50	With 25cm, inundation fills streets, parks, sidewalks, yards and parking lots. It can slow or interrupt traffic and invade simpler houses, with doors-sill next to the sidewalk level.
50	75	50cm the water will very probably have already invaded the interior of houses, causing damages to its structure and content.
75	90	At this height the water reaches practically all the goods in the interior of the houses.
>125	100	This depth has not only reached goods, but also it is enough to cause drowning.

indicates the total value of the affected properties and its contents. Where the communities suffer from inadequate sanitation, there is a higher chance of water borne diseases transmission. The normalisation of the indicator *Inadequate sanitation* is done by considering that 0% of domiciles with inadequate sanitation is the most favorable value of this parameter ($I_{IS}^{C} = 0$) whereas 100% of inadequate sanitation is the least favorable value ($I_{IS}^{C} = 100$). *Inadequate sanitation* is a variable measured by Brazilian Census.

The evaluation of the FRI concerned three areas: all the urban catchment, the fluvial corridor and the city centre, where the City Hall is located (fig. 2).

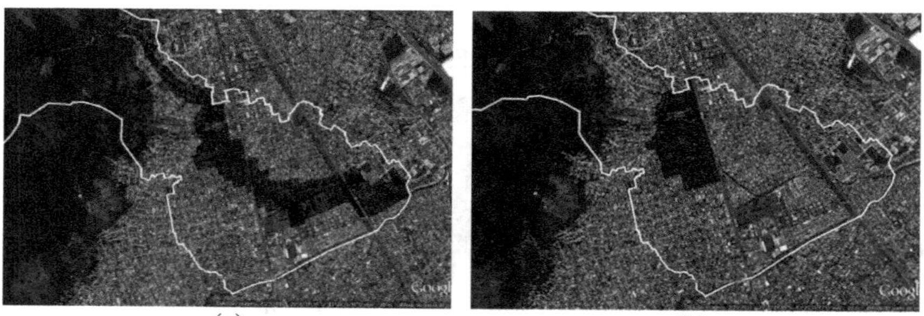

(a) (b)

Figure 2: (a) Fluvial corridor; (b) city centre [12].

The following scenarios were evaluated: the present, implementation of a dam and river restoration. The results are presented in the subsequent sections.

4.1 Present scenario

Mesquita has many infrastructure problems. In the most populated areas of the city, there are numerous illegal housing subdivisions. Even regular lots, but resulting from an intense fragmentation of the land appear frequently. Another important question is that, due to the lack of adequate housing programs, there are many houses occupying river banks (fig. 3). Rectifications and canalisations were works conducted in the majority of the rivers of Baixada Fluminense, changing a lot the lower fluvial reaches of the catchment. Dona Eugênia was one of these rivers, which had its morphology altered. Direct discharge of sewage and solid waste in the river, with visible environmental degradation is another problem. The recurrent problem of flooding affects about 80% of the population. In general terms, the degradation of the urban environment is a very critical situation, reflecting poor socio-economic aspects.

Figure 3: Illegal constructions along Dona Eugênia River (Veról [16]).

The FRI for this scenario showed the criticality of the region. For the urban part of the catchment, the obtained value was 20.6. Considering the fluvial corridor, the FRI was 23.2. For the centre part of the city, it was 34.0. The modelled flood map, for a rainfall design event of 25-year return period, is shown in fig. 4.

4.2 Dam scenario

Taking the flooding situation into account, a Water Resources Master Plan was developed, in the 1990s, for Iguaçu-Sarapuí Rivers focusing on flood control. The main propositions were traditional drainage measures, including the construction of a flood control dam at the upstream reach of Dona Eugênia River. This Master Plan was later revised, with the aim of applying sustainable urban drainage measures to the region, such as the use of urban parks, varying from the protection of the banks to temporary reservoirs. Yet, the solution for river overflows continued to depend on the design of a dam, proposed in the first study.

The dam was adjusted to the current situation (more than 10 years later) and the simulation of its operation indicated a reduction by 84% in Dona Eugênia River water levels (fig. 5). However, this fact is not replicated in the river surrounding areas, because the river banks occupation and the urban floods caused by the excess of runoff are not affected by the dam. For the urban part of the catchment, the FRI obtained value was 19.5. Considering the fluvial corridor, it was 19.2; in the centre part of the city, it was 27.4.

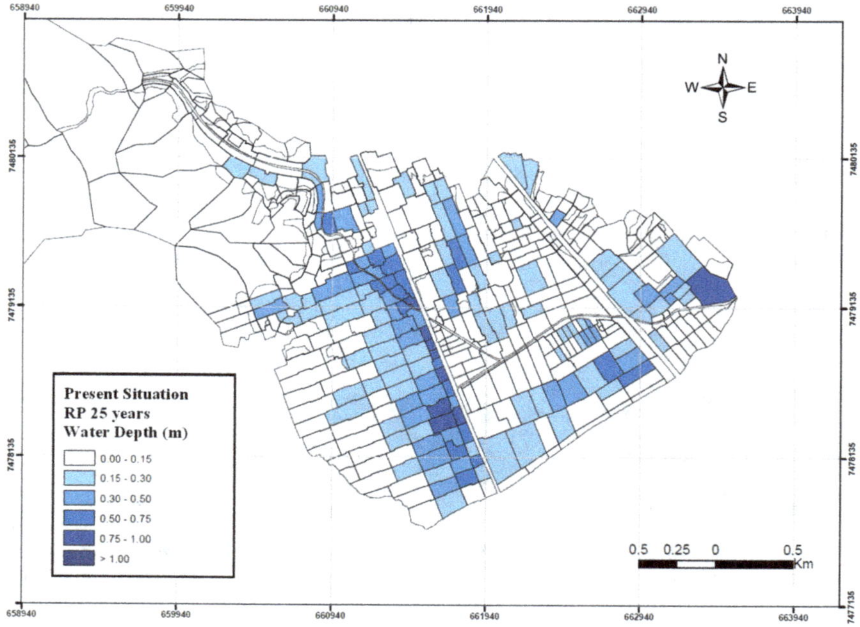

Figure 4: Flood map – present situation – RP 25 years.

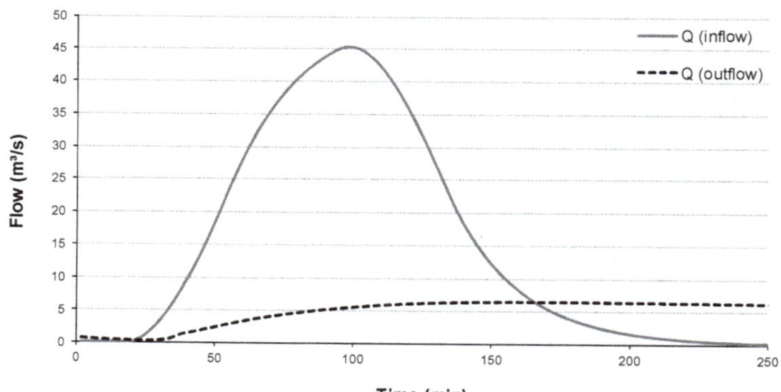

Figure 5: Hydrograph – flood control dam – RP 25 years (INEA [17]).

4.3 River restoration scenario

A river restoration (RR), although even partially, was proposed, considering new concepts of sustainable urban drainage, urban planning and development actions. The aim of this proposition was to complement or replace past development and traditional solutions by new concepts which consider fluvial ecosystems needs. Thus, four different approaches were proposed for the river treatment (as shown in fig. 6).

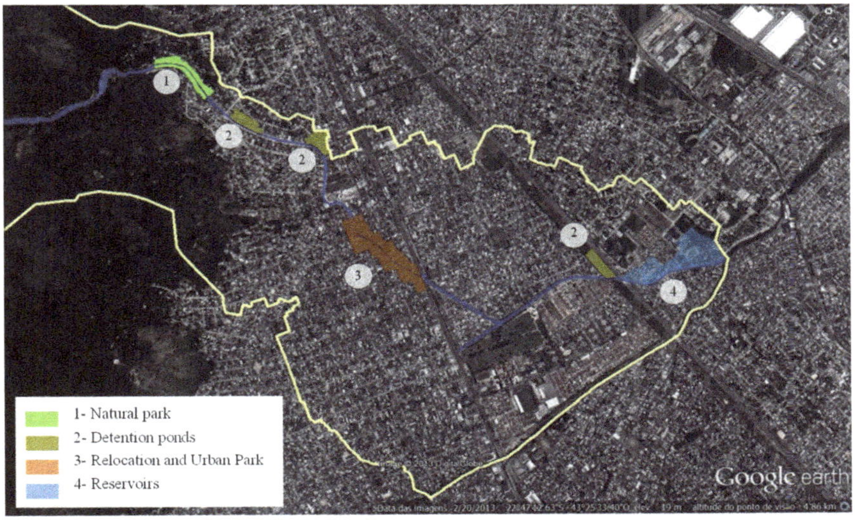

Figure 6: Areas affected by the river restoration process [12].

At the upstream reach, a more natural composition was considered, with an extension of the natural park advancing inside the city (area 1). At the downstream reach, a wetland was proposed, for temporary storage as well as a quality control measure (area 4). This last effect was not assessed - it is a possibility to be further explored. In the middle reach of the river (area 3), an urban park, with recreational purposes, was proposed to substitute the irregular occupation of the river banks (figs. 7 and 8). This area was intended to fulfil two objectives: provide an artificial flood plain for the river and give some space for relocating the population in flood proofing new buildings. Lastly, some urban interventions were introduced to act as detention reservoirs (area 2).

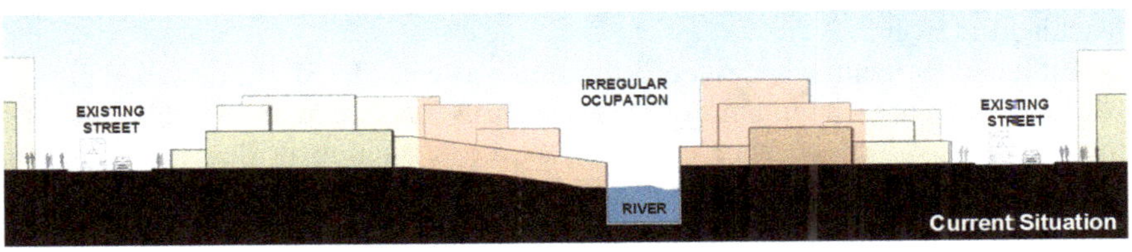

Figure 7: Section of the river in the middle part of the city [18].

In this scenario, the FRI was 18.0 for the urban part of the catchment, 12.9 when considering only the fluvial corridor and 18.6 in the centre part of the city.

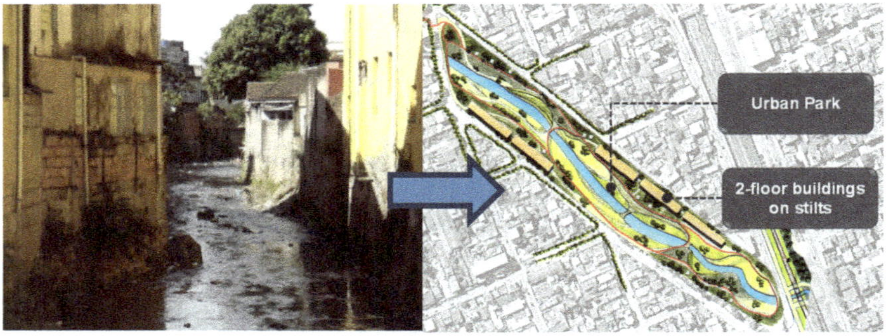

Figure 8: (a) Irregular occupation on the river banks [16]; (b) Design proposition considering an urban park and 2-floor buildings on stilts [18].

4.4 Discussion of results

As was possible to see in the present situation scenario, the flood problem in Mesquita City is critical. There are records of frequent inundation and the mathematical modelling showed important areas below water, with water depths that were able to reach, in certain places, near 1 m high, as it could be seen in the flood map of fig. 4. The centre of the city is very affected. Two important processes were verified: both major and minor drainage fail. River overflowing is responsible for flooding even the City Hall.

The proposition of a dam in the upstream reach of Dona Eugênia River was able to completely control river overflowing for a return period of 25 years. The discharges were all conducted inside the river channel. However, this measure does not act against the surface flows, the minor drainage still fails and there are several parts of the city still inundated (with minor water depths, generally). One special effect noted is in a river reach corridor, near the centre of the city, where houses irregularly placed in the river banks act as dykes (fig. 8a), preventing the connection of the river with the "flood plains". In this area, the runoff generated and not collected by the minor drainage accumulates in the streets, without reaching the river.

On the other hand, when river restoration was considered, the designed project was also capable to maintain the river within its banks, although with almost no freeboard. However, because of the possibility of acting in the riverine areas, opening spaces for temporary storage while re-connecting the river to its flood plains, part of the minor drainage problems were also solved. In the overall assessment, river restoration concepts, although limited by the built environment, produced a minor risk index, when compared with the dam alternative. Table 2 summarises the obtained results.

Table 2: FRI results.

Scenario	Urban catchment	Fluvial corridor	City centre
Present	20.6	23.2	34.0
Dam	19.5	19.2	27.4
RR	18.0	12.9	18.6

5 Concluding remarks

Few data regarding flood risk analysis or flood damage evaluation is available for this region, as for most of the Brazilian territory. However, the FRI managed to quantify flood risk by integrating various factors, expressed by easily available indicators aggregated in a simple formulation. The results that can be obtained may allow rating, identifying and comparing critical zones, as well as simulating the impact of flood control measures scenarios. The index structure is sufficiently flexible to express different flood risk perceptions, such as the decision makers', experts' or population's, by either varying the weights' or the chosen indicators' set. Considering FRI results, from the formulation used in this paper, critical flood problems in the Dona Eugênia catchment are restricted to particular and important areas, like Mesquita city centre, while most of the catchment could be rated as a low risk region. Flooding in the watershed is the result of major and minor drainage failures. The problem at those critical zones may be addressed by measures focusing either on flood hazard mitigation or on vulnerability and exposure reduction. The impacts of such measures, represented by changes in the associated indicators were forecasted with a new FRI simulation. A more traditional approach, regarding the introductions of a dam, was able to eliminate 100% of the river overflow. However, risk reduction in the city centre was something about 20%, because riverbank occupation was not addressed and there was not any effect on the superficial flows. The river restoration alternative, although only partially possible, reconnecting the flood plains to the river and freeing its banks, was also able to maintain the main flow inside the river channel and reduced flood risk in the city centre by approximately 45%. These newer techniques, associated with the concept of sustainable urban drainage, tend to treat the problem in an integrated context, focusing on distributed actions, in order to restore (as much as possible) infiltration and storage capacity of the natural basin as well as river natural hydraulic functions. The obtained results must not be seen as final ones. The river restoration scenario should be further developed, in order to introduce on-source control measures for an even more effective hydrological result. Urban development should be carried on producing low hydrologic impacts and land use occupation should also be controlled, in order to minimise river and city degradation.

Acknowledgement

The authors acknowledge CNPq for the scholarships.

References

[1] Wong, T. & Brown, R., Transitioning to water sensitive cities: ensuring resilience through a new hydro-social contract. *11th Int. Conf. on Urban Drainage*, IAHR/IWA: Edinburgh, UK, 2008.

[2] Miguez, M.G., Mascarenhas, F.C.B. & Magalhães, L.P.C., Multifunctional landscapes for urban flood control in developing countries. *International Journal of Sustainable Development and Planning*, **2(2)**, pp. 153-166, 2007.

[3] Samuels, P., Flood risk analysis and management – achieving benefits from research. *European Conference on Floods*, Vienna, Austria, 2006.

[4] Floodsite, *Integrated Flood Risk Analysis and Management Methodologies*, Available from: <http://www.floodsite.net>.

[5] Zonensein, J., Miguez, M.G., Magalhães, L.P.C., Valentin, M.G. & Mascarenhas, F.C.B., Flood risk index as an urban management tool. *11th Int. Conf. on Urban Drainage*, IAHR/IWA: Edinburgh, UK, 2008.

[6] Woods-Ballard, B., Kellagher, R., Martin, P., Bray, R. & Shaffer, P., *The SUDS Manual, CIRIA C697*, CIRIA: London, 2007.

[7] Melbourne Water. *WSUD Engineering Procedures: Stormwater*. Collingwood, Australia: CSIRO Publishing, 2005.

[8] Coffman, L.S., Cheng, M., Weinstein, N. & Clar, M., Low-impact development hydrologic analysis and design. *Proc. of the 25th Annual Conf. on Water Resources Planning and Management*, Chicago-Illinois, American Society of Civil Engineering: New York, 1998.

[9] Gusmaroli, G., Bizzi, S. & Lafratta, R., L'approccio della riqualificazione fluviale in ambito urbano: esperienze e opportunittà. *4° Convegno Nazionale di Idraulica Urbana/Acqua e Città*, Venice, Italy, 2011.

[10] Dufour, S. & Piégay, H., From the myth of a lost paradise to targeted river restoration: forget natural references and focus on human benefits. *River Research and Applications*, **25**, pp. 568-581, 2009.

[11] Findlay, S.J. & Taylor, M.P., Why rehabilitate urban river systems? *Area*, **38 (3)**, pp. 312-325, 2006.

[12] Google Earth, www.google.com/earth.

[13] IBGE – Brazilian Institute of Geography and Statistics. Demographic Census 2010. Rio de Janeiro, 2010.

[14] Mascarenhas, F.C.B. & Miguez, M.G., Urban flood control through a mathematical cell model. *Water International*, **27(2)**, pp. 208-218, 2002.

[15] Mascarenhas, F.C.B., Toda, K., Miguez, M.G. & Inoue, K., *Flood Risk Simulation*, WIT Press: Southampton and Boston, 2005.

[16] Veról, A.P., Pictures of Dona Eugênia catchment, 2012.

[17] Instituto Estadual do Ambiente (INEA), *Plano Diretor de Recursos Hídricos, Controle de Inundações e Recuperação Ambiental da Bacia do Rio Iguaçu/Sarapuí*, PDIS-RE-013-R0, vol. 2, Rio de Janeiro, RJ, 2009.

[18] Costa, I.A., Nascimento, M.B. & Furtado, V.M., Revitalização do rio Dona Eugênia: o estudo de caso da cidade de Mesquita. *Jornada de Iniciação Científica da UFRJ. Rio de Janeiro*, 2011.

Case study: flood impact reduction

Y. Peddemors & B. Willemse
Provincie Zeeland, The Netherlands

Abstract

Multi-level safety (MLS) is part of the National Waterplan (Stumpe, J., *Nationaal Waterplan*, www.nationaalwaterplan.nl), the official government plan for national water policy in the Netherlands. MLS pursues sustainable flood risk management in 2016. The MLS-approach consists of three layers. The first layer concerns prevention: it aims to minimize the occurrence of floods. The second layer aims at flood impact reduction. The third layer deals with a better preparation for the occurrence of a flood. The developed method is an area-specific risk approach that looks at an optimal combination of measures from all three layers. In this paper, the focus is on the second layer, the mandate of the Dutch provinces. Over the centuries, floods have invaded the Dutch delta area. Since the year 1200, these floods and subsequent land reclamations have formed the province of Zeeland. Many former sea dikes became inland dikes, so called "sleeper dikes". And while they have lost their original function as primary flood protection, they remain as a network of over a thousand miles of elevated tracks that can serve to reduce the impact of a flood. This makes Zeeland a region particularly suitable for MLS. To explore the (im)possibilities of these sleeper dikes, the focus has to be broadened. Due to the vast anthropogenic elements in today's world, it is nearly impossible to neglect any field of activity. Inspired by Carl Steinitz, the method of changing geography by Geodesign has been applied to the development of MLS. The objective is to achieve better spatial planning and to improve the decision-making process.

Keywords: multi-level safety, flood impact reduction, Geodesign, spatial planning.

1 Introduction

In 1953, a major storm caused enormous floods in England, Belgium and the Netherlands. In the Dutch delta area, 1836 people lost their lives and

500,000 acres of the inhabited area were invaded by the North Sea [2]. The flood of 1953 led to faster decision-making and more decisive flood protection measures. It was the direct occasion of the implementation of the Deltaworks [3].

Nowadays, preventive measures for large-scale maintenance and improvement of dikes, dunes and beaches are thoroughly embedded in the Dutch legislation. In the year 2016, reinforcement of all the dikes along the Zeeland estuaries of the Oosterschelde and Westerschelde will be completed. Meanwhile the EU Flood Directive [4] requires Member States to identify the possible impact of future floods. The province of Zeeland has calculated their future flood impacts, conform to nation-wide established (hydraulic) conditions, formulas, assumptions and manuals. The results of these flood impact calculations are open to the public since December 2013. The accompanying flood risk management plans should be ready before 2016.

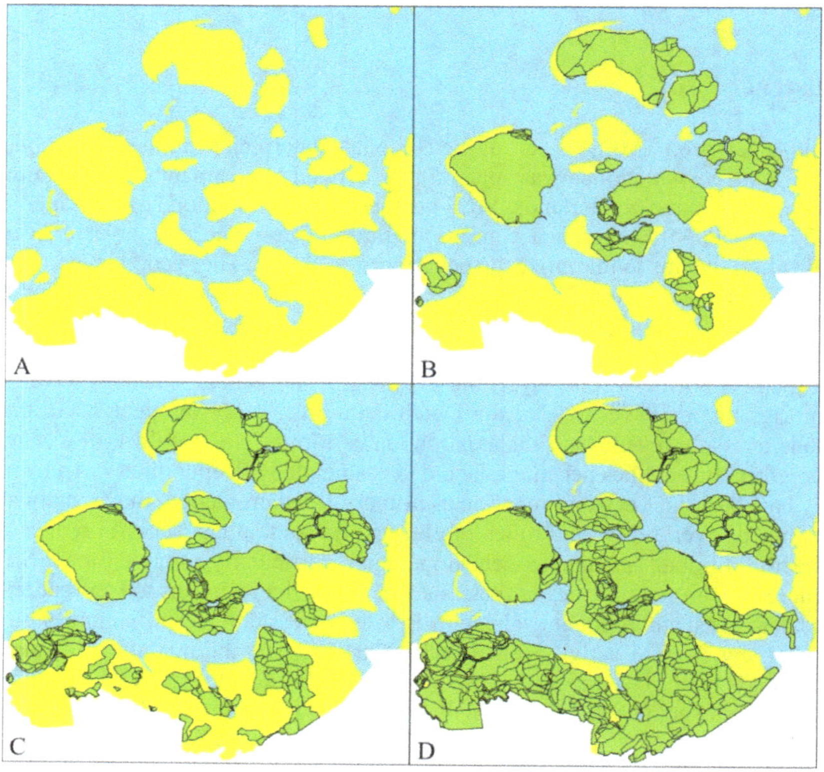

Figure 1: A – 1250; B – 1400; C – 1600; D – 1952.

The province of Zeeland is situated in the Dutch delta area. Due to area-specific geographical features Zeeland is different from the rest of the Netherlands and Northwest Europe. Since the year 1200, major floods and subsequent land reclamations have formed this delta area. Many former sea dikes became inland dikes, so called "sleeper dikes", fig. 1.

One of the nationwide established assumptions is: sleeper dikes have enough height and stability to limit the impact of floods. This assumption is made in order to compare flood impact calculations in an unambiguous way across provincial borders. However, it is not known whether this assumption is correct. More importantly, nowhere in the Netherlands this assumption has so much influence on the outcome of flood impact calculations as in a geographically unique landscape as the Zeeland estuary.

In 2016, the province of Zeeland expects to finish its research on the predictability of future flood impacts in relation to sleeper dikes. Prior to this research, several strategies have been examined in a case study entitled "flood impact reduction" involving multi-level safety (MLS) [5] and Geodesign [6].

2 Flood impact reduction

In order to achieve flood impact reduction, the province of Zeeland started to identify promising locations, elaborate integrated provincial policy and visualize spatial developments. Taking into account all sleeper dikes, this means that the focus must be on area-specific research and the aim at the predictability of future flood impacts.

2.1 Multi-level safety

Multi-level safety (MLS) is part of the National Waterplan, the official government plan for national water policy in the Netherlands. MLS pursues sustainable flood risk management in 2016.

The MLS-approach consists of three layers. The first layer concerns prevention, that is, it aims to minimize the occurrence of floods. The second layer aims at flood impact reduction. The third layer deals with a better preparation for the occurrence of a flood. The developed method is an area-specific risk approach that looks at an optimal combination of measures from all three layers. The focus in this paper is on the second layer, for which the Dutch provinces are primarily responsible.

2.1.1 Responsibilities of the province of Zeeland

The main assignment of the province of Zeeland is to develop sustainable spatial planning. This covers the second layer of MLS as well. Hence the Dutch government made the provinces responsible for calculating flood impacts

The second layer in MLS-approach also requires building customized and protecting critical infrastructure. This infrastructure supports the third layer of MLS and creates solid evacuation routes and place of safety.

2.1.2 Solutions

There are several potential solutions. The main types are related to engineering design, process management and strategy. One thing is for sure, the focus should not be on short-term planning. In the search for MLS solutions in a highly flood-prone area, each solution should have long-term vision.

2.1.3 Mission (im)possible?

The current flood impact calculations show that 90% of Zeeland will flood if the primary dikes and dunes do not withstand a storm surge. So it will be quite a difficult task to identify promising locations for sustainable spatial planning, especially considering the challenging consequence of discouraging people from living and working in flood-prone areas.

This causes the need to enlarge flood awareness in the province of Zeeland to create a wider support among citizens, businesses and decision-makers, right from the start.

2.2 Sleeper dikes

Sleeper dikes are inland dikes that at some point in earlier times separated land and sea. Over the centuries they have lost their original function as primary flood protection. In the Zeeland estuary they remain as a network of over a thousand miles of elevated tracks that can serve to reduce the impact of a flood, fig. 1.

The current flood impact calculations show that sleeper dikes have an enormous influence on the theoretical outcome of each breach in the present primary flood defences.

2.2.1 Enforcement

Maintaining strong primary flood defences, part of the first layer of MLS, is embedded in the Dutch National Waterplan. However, “waking up” sleeper dikes and reinforce some as secondary flood defences, still is under debate. Yes, sleeper dikes affect flood impacts, but can they contribute cost effectively to flood impact reduction as secondary flood defences?

Recently the province of Zeeland has identified a network of secondary flood defences that conform to the STOWA directive [7], fig. 2.

2.2.2 Secondary flood defences

The network of secondary flood defences should have a mix of the following features: hold back floods from flooding further inland, slow down and temporarily reverse floods, divert floods to less sensitive areas, serve as escape routes during a flood and serve as transport routes after a flood.

2.2.3 Assessment

To assess whether the network of secondary flood defences can make a positive contribution to flood impact reduction, the province of Zeeland should establish its own standards about required heights and stability of sleeper dikes, an assessment that only can be done experimentally.

In other words: the province of Zeeland decided to calculate additional flood impact scenarios in order to find, examine and visualize alternatives. The elaborated alternatives will show advantages and disadvantages of possible measures, specified in each diked-area.

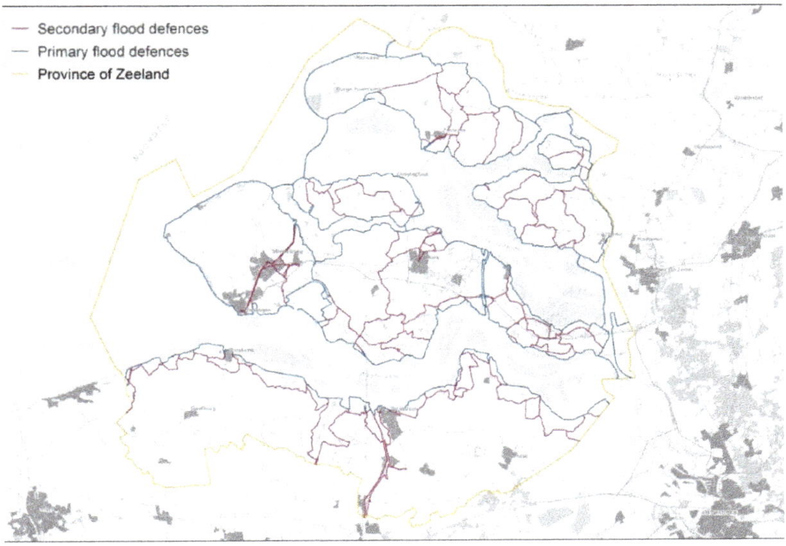

Figure 2: Secondary flood defences.

2.3 Flood impact calculations

The province of Zeeland consists of seven diked-areas: six (former) islands and the part bordering to Belgium. For each diked-area, recently calculated flood impacts assume that all elevated tracks are in the proper (hydraulic) conditions to substantially limit future flood hazards. To get a better impression of what actually happens theoretically, the province of Zeeland visualized the outcome of flood impact calculations in the current situation in stereo 3D, fig. 3.

Figure 3: Flood impacts in stereo 3D.

2.3.1 The key question

Studying the stereo 3D visualizations, the first thing to consider is whether these so called "proper conditions" are actually desirable for all our secondary flood defences and other sleeper dikes. Therefore, the province of Zeeland decided to adjust recently calculated flood impacts.

2.3.2 Adjustments

To demonstrate the desirability of the presence or the absence of sleeper dikes, the presumable sensitivity has to be identified.

The first set of additional flood impact calculations focuses on the network of secondary flood defences. Since reinforcement of sleeper dikes has never occurred, it is not realistic to assume that over a thousand miles of these sleeper dikes will remain standing during a flood. But it is a safe assumption that the network of secondary flood defences will. These 380 miles in total are strong enough not to collapse during a flood. The height of these dikes have been measured in 2012 and will be included in the simulation model. All other sleeper dikes will be erased in the simulation model.

In this case, the adjustments will be made in each diked-area. The results will be compared with the current flood impact calculations and prove the value and necessity of the presence or the absence of sleeper dikes without flood defence features. They also tell us in which diked-area the influence of the nation-wide assumption is the most sensitive.

2.3.3 Promising locations

The second set of additional flood impact calculations focuses on the promising locations. Area-specific features have to be established. It needs to be examined which of the five features of the secondary flood defences, as mentioned in paragraph 2.2.2, is the most important.

This priority ranking cannot be imposed by the government, but must be developed in consultation with all regional stakeholders.

3 Case study

Zeeland contains 7 diked areas, each with its own history, dedicated spatial planning and scale. This requires a different policy for each diked area. In order to obtain a kind of similarity, a process approach is necessary.

3.1 Geodesign

Geodesign is changing geography by design, a decision driven method designed by Steinitz [6]. It provides an iterative framework with loops and feedback. There are six questions for each step to answer, namely the following questions: why, how, what, where and when, as they appear in fig. 4. This process requires collaboration between the design professionals, geographic sciences, information technologies and the local people.

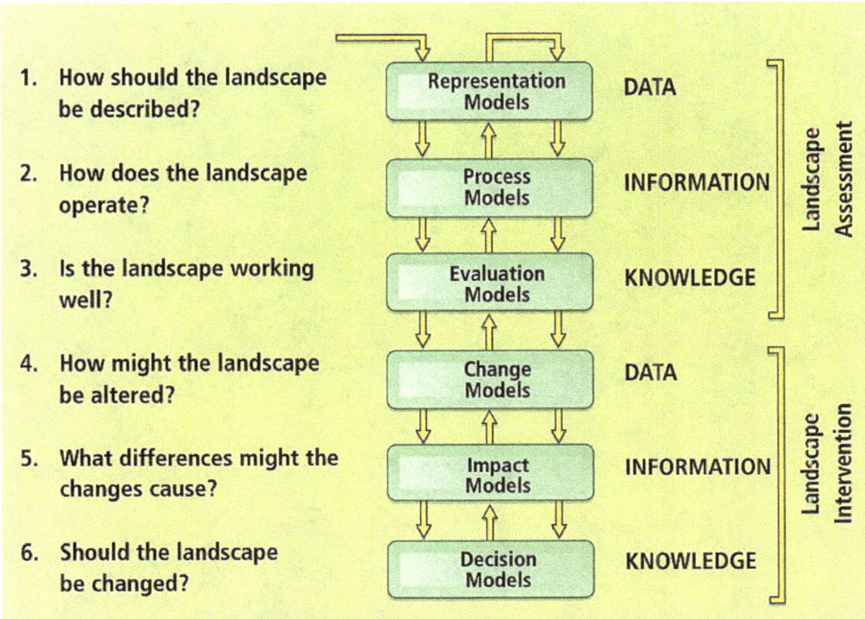

Figure 4: Framework for Geodesign.

3.2 Why Geodesign

Due to the vast anthropogenic elements in today's world, it is nearly impossible to neglect any field of activity. Humanity has granted the earth with a very complex layer. An integrated approach is needed to combine design professions and geographic sciences. By applying a method built on collaboration and cooperation, it is most likely that insights are included.

"Geodesign lies within the domain of applied science and engineering, seeking ways of addressing practical problems using the scientific method" [8].

The objective is to enable better spatial planning, to improve communication for decision-making and gain experience by conducting case studies.

3.3 Case studies

Early in 2014, two trial sessions with Geodesign were organized. The first group consisted of students and people of the place, the second consisted of professionals of the province of Zeeland. Both groups were informed on the method and provided with several maps (satellite images, false colour, height, topographic, flood water speed, flood water front) (as shown in fig. 5).

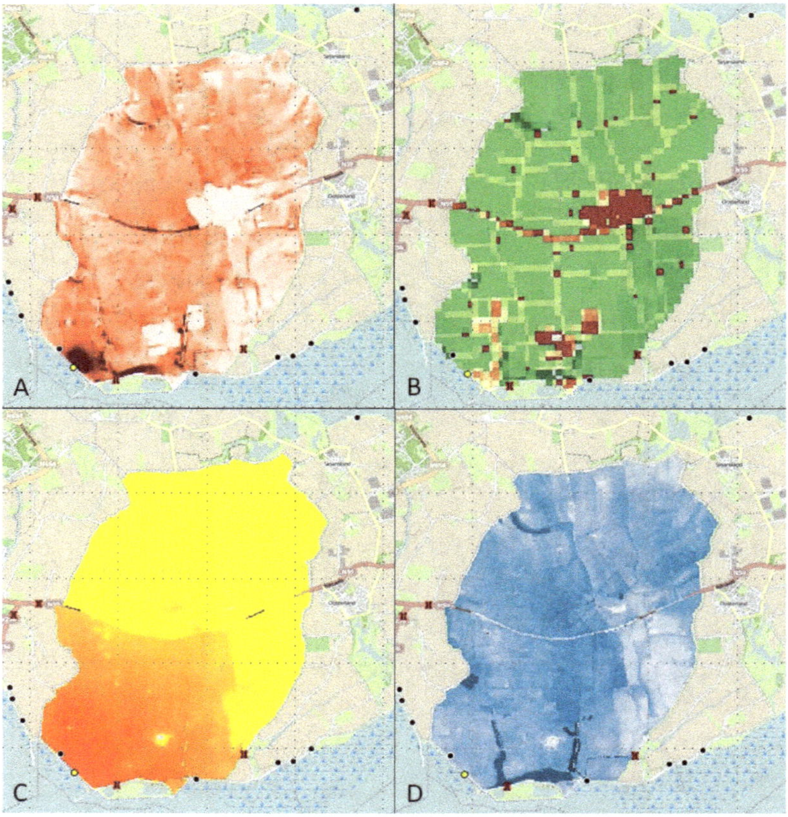

Figure 5: A – Velocity; B – damage; C – arrival; D – max. water depth.

3.4 Results

Since time was limited, some information was missed: information on economy, vital infrastructure, traffic and population numbers. During the trial sessions of the province of Zeeland, both test groups came up with a rather challenging statement: "We actually do have to discourage building in flood-prone areas".

3.5 Research proposal

In March 2014, the province of Zeeland joined forces with Associate Professor Dr. Stan Geertman in a proposal for doctoral research entitled "A map-based Planning Support System (PSS) for land and water governance". The overall question of this programme is: What are the required functionalities of a Planning Support System that supports water managers and spatial planners to make decisions for flood risk management under changing governance conditions. The province of Zeeland will participate in conducting the case studies in Zeeland.

Figure 6: Geodesign in progress.

4 Conclusion

During trial sessions, it has been learned that Geodesign is an effective method in identifying promising locations in Zeeland for the application of the MLS-approach. To elaborate an integrated regional vision, the province of Zeeland starts using Geodesign in spatial planning to visualize alternatives in flood impact reduction.

References

[1] Stumpe, J., *Nationaal Waterplan*, www.nationaalwaterplan.nl, Dutch Government, Thieme Deventer, 22 December 2009.

[2] Slager, K., *Watersnood*, ISBN 978.90.71359.13.2, Joh. Enschedé, Amsterdam, 2010.

[3] Stichting Deltawerken online, www.deltawerken.com

[4] European Commission Directive 2007/60/EC, http://ec.europa.eu/environment/water/flood_risk/, 26 November 2007.

[5] Kolen, B., Zethof, M. & Maaskant, B., *Toepassing basisvisie afwegingskader meerlaagse veiligheid; een methode om mee te werken in de praktijk*, ISBN 978.90.5773.554.7, STOWA, 2012.

[6] Steinitz, C., *A framework for Geodesign* CA: ESRI Press, 2012.

[7] Geerse, C., Stijnen, J. & Kolen, B., *Richtlijn normering compartimenteringskeringen*, ISBN 978.5773.408.3, STOWA, December 2007.

[8] Goodchild, M.F., Towards Geodesign: repurposing cartography and GIS? *Cartographic Perspectives*, **(66)**, pp. 7-21, 2010.

Assessment of the flood disaster management plans for the medical services in Tokyo and Fukuoka, Japan

M. Takezawa[1], H. Gotoh[1], K. Suzuki[1], Y. Kakehi[2] & T. Yamamoto[3]
[1]*Nihon University, Japan*
[2]*Fukuoka City, Japan*
[3]*Chuo College of Technology, Japan*

Abstract

Recently, large scale flooding disasters due to unexpected weather have been reported worldwide. These flooding disasters have caused severe damage and have claimed many lives. The medical services play an important role in flood disasters. In this paper, a disaster management plan for the medical services is described in terms of governmental policy, utilising questionnaire survey results, among others. The target areas used for the survey are the lowlands from the former flood disaster areas in Tokyo and Fukuoka, Japan. The policy of the Japanese Government regarding large scale flood hazards has been discussed by a team of experts. Moreover, some flood hazard prevention methods were proposed by the disaster risk management offices of the local governments of Tokyo and Fukuoka. The survey of flood preparedness for the medical services was sent and collected by mail. It was clear from the questionnaire results that the flood hazard measures of the medical services in large cities such as Tokyo and Fukuoka are fairly poor, although the Japanese government and local governments do provide a guide for a disaster event. To reduce disaster risk, it is necessary to encourage the promotion of a mechanism that can provide guidance for the prevention of flood hazards in the building and planning of cities and towns, to establish a set of spacious shelters, to secure evacuation routes etc., within a guide that will illuminate a crisis consciousness for flood hazards.
Keywords: medical service, flood preparedness, questionnaire, flooding hazard.

1 Introduction

Flooding is a dangerous disaster that not only severely damages natural resources but also results in loss of life. Flood disasters are usually caused by heavy rainstorms that overflow bodies of water or inundate sections of land. Flood preparedness is largely dependent on two elements: first, the ability of the relevant national, local and community institutions to orchestrate communication; and second, determining and prioritizing the content of communications on the basis of user needs and priorities. The lead time of the hydrologic forecasts is very short and people do not fully comprehend danger-level terminology. There is no mechanism for relating any forecast information with users' needs at specific locations. People have different capacities and vulnerabilities in regards to the dissemination of information due to their different roles and conditions and are therefore affected differently by a disaster. In many contexts, people are better connected to the early warning mechanisms due to their movement in public spaces and access to various channels of communication, such as radio, TV, informal community networks and interaction with officials. Japan has suffered extensive damage and casualties from frequent flooding and the medical services are very important for treating the wounded and injured when disaster strikes. The primary medical services in this paper refer to the large hospital where sick or injured people receive medical treatment as inpatients and the clinic where medical treatment is provided to people as outpatients. Emergency medical services are especially important in the event of a flood disaster.

2 Past flooding disasters in Japan

Japan consists of a series of small islands making it susceptible to a number of natural disasters. Typhoons are massive storms that can ravage the islands with high winds and storm surges. Typhoon Vera that hit Ise Bay on the Pacific Ocean of the midland in Japan in July 1959 killed more than 5,000 people and left more than 1.5 million homeless. The rainy season in Japan is from May through to July. Heavy rains fed by the warm ocean waters surrounding Japan can cause significant flooding and mudslides on the islands [1]. Typhoon Kathleen struck the entire Kanto Region in Japan on September 15, 1947. Heavy rains caused the Arakawa and Tone Rivers to overflow and the resulting floods killed 6 people, left 2 people missing and 3 people injured in the Tokyo metropolitan area [2]. Typhoon Kitty struck the Tokyo/Yokohama area between August 31 and September 1, 1949. A total of 18 people were killed and the number of injured reached 104 people due to rainfall-induced flooding and landslides [3]. Typhoon Ida in 1958 was the third deadliest typhoon in Japan. Ida caused torrential flooding in southeastern Japan, resulting in over 1,900 mudslides [4]. There were a total of 36 deaths, with 3 people missing. The Kyushu district of southern Japan suffered extensive damage from flooding in 2012 [5]. At least 20 people were killed by flooding, 7 people were missing and approximately 250,000 people were ordered to evacuate. Evacuation orders affected nearly 85,000 households in the Kyushu district and school gyms were being used as evacuation centers. More than 2,600 houses were destroyed by the flooding. Heavy rainfall caused by a stationary front inundated the Kyushu district in 2003 [6]. During this heavy rainfall, 1,352 houses were flooded when the Mikasa River overflowed its banks in Fukuoka City. A flood disaster also occurred in Fukuoka City in 1999 [7]. A disaster prevention plan based on internet real-time rainfall and water level information had been established in a hotel after the disaster. Water sealing plates were installed according to this plan and this hotel escaped flood damages. This shows the importance of examining the methods for utilizing real-time rainfall information. The North Kyushu Flood hit Northern Kyushu in 1953 [8]. The torrential rain during the rainy season, amounting to more than 1,000 mm on Aso and Hikosan mountains, caused great flooding in many of the rivers, including the Chikugo River, and left more than 1,000 people either dead or missing. A total of 450,000 houses were flooded and 1,000,000 people suffered from the flood. Fundamental changes in flood control in the Kyushu area were needed and flood control standards have now been established to prevent further floods.

3 Flooding prevention and medical services in Japan

For effective disaster management, the national government, local governments and a wide range of relevant partners designated that public corporations should formulate disaster management plans and carry them out appropriately, based on the Disaster Countermeasures Basic Act [9]. To protect national land as well as citizens' lives, livelihoods and property from disasters is a national priority. The turning point for strengthening the disaster management system came after the immense damage caused by the Isewan Typhoon (Typhoon Vera) in 1959 and led to the enactment of the Disaster Countermeasures Basic Act in 1961, which established a comprehensive and strategic disaster management system. The main contents of the Disaster Countermeasures Basic Act are (1) Definition of responsibilities for disaster management, (2) Disaster management organizations, (3) Disaster management planning system, (4) Disaster prevention and preparedness, (5) Disaster emergency response, (6) Disaster recovery and rehabilitation, (7) Financial measures and (8) State of Disaster Emergency.

Japan is prone to a variety of water and wind-related disasters including flooding, landslides, tidal waves and storm hazards, owing to meteorological conditions such as typhoons, active weather-front systems and geographical conditions such as precipitous terrains and steep rivers, as well as settlement conditions, in which many of the cities are built on river plains. One-half of the population is concentrated in possible inundation areas, which account for about 10% of the national land. Although there has been a large reduction in the area inundated by floods owing to soil conservation and flood control projects over many years, the amount of general assets damaged in flooded areas has increased in recent years. Additionally, as a long-term trend, there is an increasing tendency of downpours throughout the country. In order to reduce damage caused by severe weather disasters, structural measures such as improving rivers, dams and sewage systems and non-structural measures such as preparing hazard maps and providing disaster information must be promoted in an integrated manner. As non-structural countermeasures, the warning and evacuation systems of the possible inundation areas and landslide-prone areas have been developed in accordance with the Flood Control Act and the Act on Promotion of Sediment Disaster Countermeasures for Sediment Disaster Areas. Both laws were amended in 2005 to intensify measures, including the familiarization of hazard maps and the identification of a method to disseminate disaster information to facilities caring for those who require assistance at the time of a disaster, such as elderly people, in the municipal disaster management plans. In light of a rising trend in heavy downpours in recent years, there is a strong need to fortify measures for rapid, effective evacuation and relief, in anticipation of large-scale flood disasters.

Emergency hospitals are a very important medical service in times of large-scale flood disaster. Emergency medical services are services dedicated to providing out-of-hospital acute medical care, transport to definitive care and other medical transport for patients with illnesses and injuries that prevent patients from transporting themselves. The emergency medical services may also be locally known as paramedic services, first aid squads, emergency squads, rescue squads, ambulances, squad ambulance services, ambulance corps or life squads. The goal of most emergency medical services is to either provide treatment to those in need of urgent medical care, with the goal of satisfactorily treating the presenting conditions, or arranging for timely removal of the patient to the next point of definitive care. This is most likely an emergency department at a hospital. The term “emergency medical service” has evolved to reflect the change from a simple system of ambulances providing only transportation, to a system in which, actual medical care is given on scene and during transport. In some developing regions, this term is not used, or may be used inaccurately, since the service in question does not provide treatment to the patients, but only the provision of transport to the point of care. In most places in the world, the emergency medical service is summoned by members of the public (or other emergency services, businesses or authorities) via an emergency telephone number that puts them in contact with a control facility, which will then dispatch a suitable resource to deal with the situation. In some parts of the world, the emergency medical service also encompasses the role of moving patients from one medical facility to an alternative one; usually to facilitate the provision of a higher level or more specialized field of care but also to transfer patients from a specialized facility to a local hospital or nursing home when they no longer require the services of the specialized hospital, such as following successful cardiac catheterization due to a heart attack. For such services, the emergency medical service is not summoned by members of the public but by clinical professionals (e.g. physicians or nurses) in the referring facility. Specialized hospitals that provide higher levels of care may include services such as a Neonatal Intensive Care Unit (NICU), Pediatric Intensive Care Unit (PICU), state regional burn centers, specialized care for spinal injury and/or neurosurgery, regional stroke centers, specialized cardiac care (cardiac catheterization) and specialized/regional trauma care. In some jurisdictions, emergency medical service units may handle technical rescue operations such as extrication, water rescue and search and rescue. Training and qualification levels for members and employees of emergency medical services vary widely throughout the world. In some systems, members may be present who are qualified only to drive the ambulance, with no medical training. In contrast, most systems have personnel who retain at least basic first aid certifications, such as Basic Life Support (BLS). Additionally, many emergency medical service systems are staffed with Advanced Life Support (ALS) personnel, including paramedics, nurses, or, less commonly, physicians. In order to make it possible to promptly and appropriately provide medical care to the injured, etc. when a large disaster occurs, the first medical care system of the Disaster Medical Assistance Team and medical aid unit, etc., the injured person transporting system, and the information exchanging system have been upgraded, with the focus on the disaster medical care coordinators. The Disaster Medical Assistance Team comprises specially trained medical teams that are dispatched to disaster locations, such as natural disasters, to provide emergency medical care as quickly as possible. Damage estimates for an earthquake with an epicenter in the Tokyo metropolitan area predict extensive injuries and loss of life. To ensure the effectiveness of rescue efforts through maintaining partnerships with each of the Tokyo Metropolitan Government disaster prevention organizations, the “Tokyo Metropolitan Government Regional Disaster Prevention Plan” establishes medical care systems during a disaster, including medical information gathering, initial medical care (dispatch of Tokyo Disaster Medical Assistant Team, triage), transportation of casualties and backup hospitals (disaster base hospitals). The disaster medical care steps are shown in fig. 1 [10, 11].

The Tokyo Disaster Medical Assistance Team is a group of medical personnel that are dispatched at the time of a disaster, such as a large earthquake or a large-scale traffic accident, to administer emergency medical treatment at the scene. In principle, each team has one doctor and two nurses. The Tokyo Disaster Medical Assistance Team made it a goal to have 22 hospitals and 700 members to cover each of the 12 medical care zones in Tokyo within the fiscal year 2006. During a disaster, it is important to effectively distribute limited medical resources, personnel and medicine to treat as many victims as possible. In line with that purpose, within the triage system, patients are treated or transported depending on the severity and urgency of their injuries. The triage criteria are shown in table 1.

After triage, casualties will be taken to medical facilities by the Tokyo Fire Department transport, medical care team vehicles, helicopter, boat or some other form of transport. The Tokyo Metropolitan Government designates 70 hospitals in the Tokyo metro area as “Tokyo Disaster Base Hospitals” to expand and strengthen the backup hospital system during a disaster. These hospitals may display the designation of “Tokyo Disaster Base Hospitals” at their entrances. General hospitals will also make open beds available and temporarily enhance capacity as the disaster base hospitals cannot house all casualties. Because injuries during a disaster range from minor to severe, as a general rule doctors triage from the entrance of the medical care station.

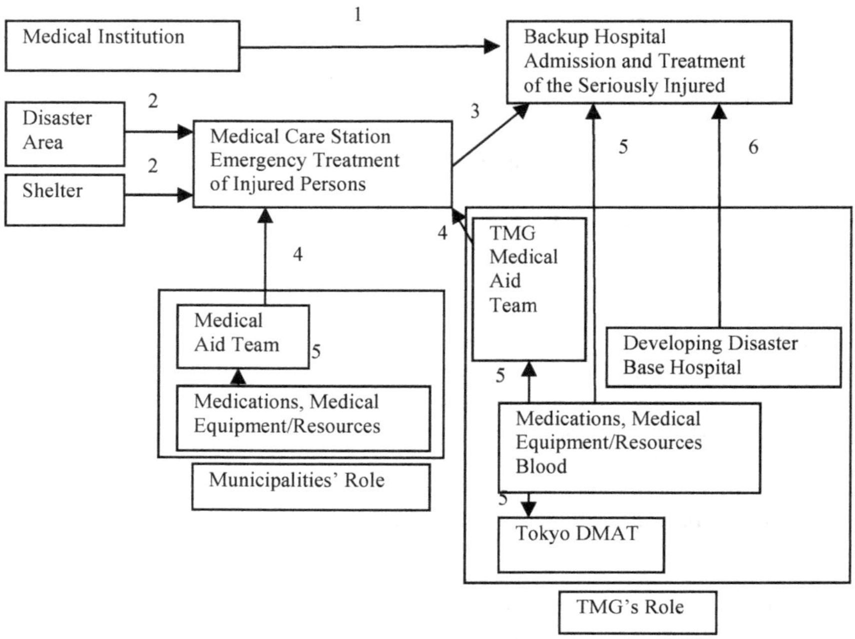

Figure 1: Disaster medical care steps [10]. 1: Person who receives limited medical care (this means someone who cannot receive medical care because medical institutions have been damaged). 2: Injured persons. 3: Serious cases. 4: Dispatch. 5: Supply. 6: Development.

Table 1: Triage criteria [10].

Categories	Priority	ID Tag	Symptoms, etc.
Immediate (Severe)	First	Red	Requires immediate treatment to save life. Symptoms include asphyxia, excessive bleeding, state of shock.
Delayed (Moderate)	Second	Yellow	(1) Can have their treatment delayed and the condition is not life threatening. (2) Principally, when vital signs are stable.
Wait (Minor)	Third	Green	Those with minor injuries other than those outlined above. Does not require a specialist's care.
Deceased	Fourth	Black	Confirmed or obviously deceased. Unlikely to be revived given cardiopulmonary resuscitation.

4 Questionnaire results

The questionnaire was sent to 300 medical services of four wards (Katsushika, Edogawa, Sumida and Kotoh) along the Arakawa River and the Sumidagawa

River in the Tokyo metropolitan area and seven wards (Higashi, Hakata, Cyuou, Minami, Jyonan, Sagara and Nishi) along the Nakagawa River and the Mikasagawa River in Fukuoka City, as shown in figs. 2 and 3. The Arakawa River is one of the principal rivers flowing through downtown Tokyo. The Arakawa River originates at Mt. Kobushi in Saitama Prefecture and empties into Tokyo Bay [12]. The Sumidagawa River flows through downtown Tokyo. The Sumidagawa River branches from the Arakawa River at Iwabuchi and flows into Tokyo Bay [13]. The Nakagawa River originates at Mt. Seburi and flows into Hakata Bay through downtown Fukuoka City [14]. The Mikasagawa River originates at Mt. Houman and flows into Hakata Bay joining some rivers [15].

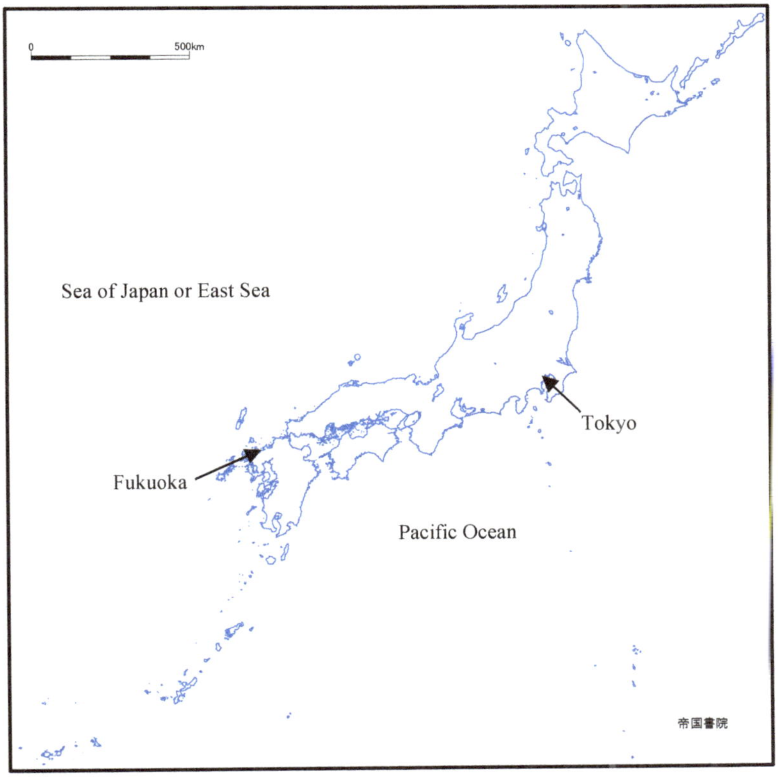

Figure 2: Map of Japan [16].

The questions used within the questionnaire are as follows:

(1) Do you have facilities to accept patients for treatment? (*yes* or *no*)
(2) If the answer to (1) is *yes*, then how many beds do you have?
(3) How long has your medical service been active?
(4) What floors are the MRI, CT, X-ray centers?

(5) Does your medical service have auto-generators? (*yes* or *no*)
(6) If the answer to (5) is *yes*, how many stories are the auto-generators placed at?
(7) Has your medical service been damaged by flooding? (*yes* or *no*)
(8) Does your medical service have a flooding manual? (*yes* or *no*)
(9) Are you referring to following the flooding hazard maps provided by the Civil Service for evacuation? (*yes* or *no*)
(10) Do you think that the flooding management of your medical service is sufficient? (*yes* or *no*)
(11) If the answer to (10) is *no*, what are the specific flooding measures of your medical service?
(12) Does your medical service stockpile food and supplies? (*yes* or *no*)
(13) If the answer to (12) is *yes*, how much is stockpiled?
(14) If the answer to (12) is *yes*, how many stories does your medical service stockpile their food and supplies?
(15) Does your medical service stockpile medicines? (*yes* or *no*)
(16) If the answer to (15) is *yes*, how many medicines are stockpiled?
(17) If the answer to (15) is *yes*, how many stories does your medical service stockpile their medicines?
(18) Does your medical service hold flooding evacuation drills? (*yes* or *no*)
(19) Does your medical service provide an evacuation boat during flooding? (*yes* or *no*)
(20) Is your medical service going to start with how timing of refuge for the flooding?
(21) Is the evacuation route secured? (*yes* or *no*)
(22) If the answer to (21) is *no*, what is the specific evacuation route?
(23) What other opinions do you have regarding the flooding preparedness measures of your medical services?

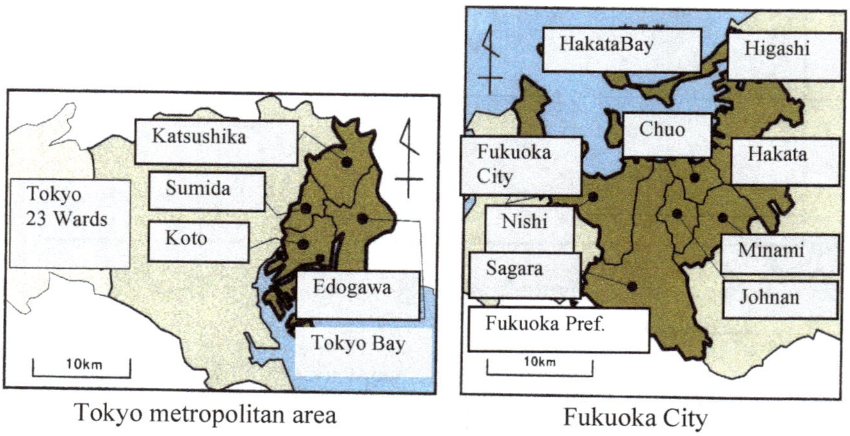

Figure 3: Locations of Tokyo metropolitan area and Fukuoka City.

The results of the above questionnaire are given in table 2.

Table 2: Results of questionnaire (%).

No	Yes of T	No of T	? of T	Yes of F	No of F	? of F
1	30.1	69.9	0	20.6	79.4	0
2	Average			Average		
3	Average 27 years			Average 20 years		
4	1st 75% 2nd over 25%			1st :65% 2nd over 35%		
5	34.2	64.4	1.4	19.0	81.0	0
6	5th under: 35% 6th over: 65%			5th under: 40% 6th over: 35%		
7	2.7	95.9	1.4	7.9	92.1	0
8	9.6	90.4	0	9.5	90.5	0
9	42.5	53.4	4.1	36.5	63.5	0
10	4.1	87.7	8.2	14.2	81.0	4.8
11	Placement of sandbags and cutoff plates			Raise the floor over the supposed inundation depth		
12	45.2	54.8	0	27.0	69.8	3.2
13	172 meals			202 meals		
14	Almost all on 1st or 2nd floor			Almost all on 1st or 2nd floor		
15	60.3	35.6	4.1	52.4	41.3	6.3
16	7 days/person			9 days/person		
17	Almost 1st or 2nd floor			Almost 1st or 2nd floor		
18	1.4	97.2	1.4	7.9	90.5	1.6
19	0	98.6	1.4	0	96.8	3.2
20	Personal judgment			Follow the evacuation advice of the government		
21	34.3	57.5	8.2	39.7	49.2	11.1
22	Narrow aisle for evacuation			No place for evacuation		
23	Money for remodeling the upper floors Simple manual pertaining to the region Specific indication for hazard			Check and maintenance of drainage Raising awareness of hazard maps Specific indication for dangerous places		

T: Tokyo Metropolitan, F: Fukuoka City.

5 Discussions with management

There was flooding in the vicinity of the junction between the Uonogawa River and the Shinanogawa River in 2011. As a result, the 1st floor of the Aorinosato special nursing home for the aged (shown in fig. 4), was flooded. For future reference, a hearing was conducted on the management of the home to discuss the situation during the flooding. (1) The location of the special nursing home for the aged was selected by considering the medical service, the welfare and the education for residents. (2) The hazard information was acquired by the crisis of home staff. (3) A manual was not prepared for flooding disasters, although one was prepared for earthquakes. Moreover, the flooding evacuation route was not prepared, although one was prepared in the case of a fire. (4) The food and the medicine were kept on the 2nd floor or above. (5) The total amount of damage was $90,000, but was not fully compensated because of the high cost of some of the instruments. (6) The reception room was carefully disinfected using antiseptics after the flood hazard. (7) The air conditioning was very important. It was clear that the home staff were able to meet the challenges posed by the flooding. In addition, they must endeavor to carry out retrieval at the home. The medical service should correspond to flooding hazards via the hazard manual as conducting fair action in the case of an emergency.

Figure 4: Aorinosato special nursing home for the aged [17].

6 Conclusions

The flood preparedness of the medical services was considered using the results from the questionnaire and discussion. The main findings are as follows:

(1) Government grants are necessary for preparing flooding defense measures.
(2) Convenient storage facilities for sectional water-barriers and defenses should be secured in the surrounding site.
(3) Drainage channels are constructed and repaired using gutters, reservoirs and cisterns.
(4) Emergency food, supplies and medicines are to be kept on the floor above the maximum inundation water depth of flooding.

(5) High-cost medical equipment, such as MRI, CT and X-ray machines are to be placed on the highest floor possible, or their centers are to be constructed in waterproof rooms.
(6) Medical services are to set up auto-generators and air conditioners in consideration of long-term stays due to the inundation.
(7) The Japanese Government, the Tokyo metropolitan area and Fukuoka City are to ensure that medical services are practiced on a place higher than the predicted inundation depth.
(8) Members of medical services should maintain crisis consciousness for flooding and also prepare a manual of flooding disaster measures.
(9) It is necessary to perform flood evacuation drills.
(10) The Disaster Imagination Game can be used to raise the awareness of evacuation in the event of flooding for those that have difficulty walking.
(11) It is necessary to consider duly and discuss seriously among the civil service, the medical services and inhabitants how to transfer them from the devastated area to a safety area, and how to secure both the transport route and the transportation. Moreover, it is also important to construct a disaster center of placing on the access in the assignment of refuge buildings in cooperation with the neighboring facilities like schools, offices, corporation, etc. by the guidance of the government.

It is clear from the above results that the disaster prevention by medical services in downtown Tokyo and Fukuoka is incomplete at present. In the future, it will be necessary to improve the disaster prevention by medical services, with focus on small children and senior citizens. Furthermore, the government should guide the city plans to strengthen measures against disasters and provide funding towards improving the medical services in the event of future large-scale disasters.

References

[1] Wikipedia, the free encyclopedia, Typhoon Vera, http://en.wikipedia.org/wiki/Typhoon_Vera, 17/02/2014.

[2] Wikipedia, the free encyclopedia, Typhoon Kathleen, http://ja.wikipedia.org/wiki/%E3%82%AB%82%B9%E3%83%AA%E3%83%B, 2014, (in Japanese).

[3] Wikipdia, the free encyclopedia, Typhoon Kitty, http://ja.wikipedia.org/wiki/%E3%82%AD%E3%83%86%E3%82%A3%E5%8F%B0, 2014 (in Japanese).

[4] Wikipedia, the free encyclopedia, Typhoon Ido, http://en.wikipedia.ord/wiki/Typhoon_Ida_(1958), 2014.

[5] The Japan Daily Press, "Kyushu rains, flooding subsides as death toll rises to 26", http://japandailypress.com/kyushu-rains-flooding-subsides-as death-toll-rises-to-28-1 2014.

[6] Wikipedia, the free encyclopedia, Typhoon Soudelor (2003), http://en.wikipedia.org/wiki/Typhoon_Soudelor, 17/02/2014.

[7] Kusuda, T., The heavy rain in Fukuoka City. *Japan Society of Civil Engineers*, **84**, 1999.

[8] Wikipedia, the free encyclopedia, 1953 North Kyushu Flood, http://en.wikipedia.org/wiki/1953_North_Kyushu_Flood, 12/02/2014.
[9] Cabinet Office, Government of Japan, *Disaster Management In Japan*, 2011
[10] Tokyo Metropolitan Government, *Disaster Prevention Guide Book*, http://www.bousai.metro.tokyo.jp/, 17/02/2014.
[11] Asia News Networks, Disaster base hospital in Japan fear flooding, http://www.asianewsnet.net/nws-40586.html.
[12] Wikipedia, the free encyclopedia, Arakawa (Kanto), http://ja.wikipedia.org/wiki/%E8%8D%92%E5%B7%9D., 18/02/2014 (in Japanese).
[13] Wikipedia, the free encyclopedia, Sumida River, http://en.wikipedia.org/wiki/Sumida_River, 18/02/2014.
[14] Wikipedia, the free encyclopedia, Nakagawa, http://ja.wikipedia.org/ wiki/%E9%82%A3%E7%8F%82%E5%B7%9D., 18/02/2014 (in Japanese).
[15] Wikipedia, the free encyclopedia, Mikasagawa, http://ja.wikipedia.org/wiki/%E5%BE%A1%E7%AC%A0%E5%B7%9D, 18/02/2014 (in Japanese).
[16] Teikoku Shoin, White Map in Japan, http://www.teikokushoin.cojp/howto/outline_map/japan/.
[17] Social Welfare Service Corporation Ojiya Kitauonuma Welfare Society, Aorinosato, http://ok-fukushikai.com/index.php?id=6 (in Japanese).

Determination of the weighting factors of criteria influencing highway flooding using multi criteria analysis based on an analytic hierarchy process

S. Budhakooncharoen[1] & B. Dhabhisara[2]
[1]*Mahanakorn University of Technology, Thailand*
[2]*Chulalongkorn University, Thailand*

Abstract

Flooding is one of the major natural disasters affecting millions of people. Thailand is also frequently faced with this type of disaster. In particular, the 2011 mega flood in Central Thailand, which inundated highways, severely contributed to the failure of national economy and increased the risk to life. Lessons learned from such an extreme event encouraged flood monitoring and warning, two basic elements of sound mitigation. Weighting factors for each cause of highway flooding are considered as essential information for classifying the degree of highway flood hazard and to enable flood monitoring and flood warning in hazardous areas. The causes of highway flooding have been investigated and it has been found that the three major factors influencing highway flooding are the characteristics of the highway, the hydrology and the land topography. The importance of each cause of highway flooding in the whole country was assessed by weighting these three major factors using multi-criteria analysis (MCA) based on the analytic hierarchy process (AHP). The MCA, based on the AHP, was selected due to its being a structured technique for organizing and analyzing complex decisions based on mathematics and psychology. According to the results of MCA, the highway, hydrological and topographic characteristics of the land were weighted as 35, 35 and 30%, respectively, with regard to influencing or causing highway floods. These weighting factors are estimated values indicating the relative importance or impact of each highway flood cause in a whole as compared to each other. The methodology of weighting factors determination will be clearly presented in this paper. These factors are applicable for setting the

priority for working out how to fix the problems and help reduce flood incidents. In addition, these factors will provide essential information for further study in classifying the degree of flood for highway flood hazard mapping.
Keywords: analytic hierarchy process, flood hazard, flood mitigation, Thailand flood, multi criteria analysis.

1 Introduction

Global warming in recent decades and its projected continuation is the major concern of various human sectors. Over the past decades, many urban areas in the world have been experiencing a dramatic increase in natural disasters due to water-related extreme events. The lack of precipitation as well as heavy downpour causes disasters ranging from extreme drought to unprecedented flood. In addition, climate change, environmental degradation, population growth, urbanization and anthropogenic effects cause human society to be more vulnerable to water-related disasters.

Flooding is one of the major natural disasters affecting millions of people, destroying property and resulting in loss of human life. Nowadays, the world is facing an increasing number of floods, not only in inundated areas but also in their frequency. Thailand, in particular, is frequently faced with this type of disaster – especially the 2011 mega flood in the central area of the country, the Chao Phraya river basin, which severely contributed to the failure of industry and agriculture.

Although the residents in the Chao Phraya and related basins have learnt to adapt their lifestyle to dealing with the recurrent flood cycle by investigating the variation of the rain, water stage and surrounding environments, the 2011 mega flood in Central Thailand is one of the hardest cases. Highway flooding in that year severely contributed to the failure of national economies and increased the risk to life. Lessons learned from such an extreme event encouraged flood monitoring and warning; two basic elements of sound mitigation. This is because highway flooding is a potential risk to life and limb and can potentially damage property. Flood monitoring and warning help mitigate loss of lives and property losses.

Although the flood hazard mapping was successfully accomplished in some areas in Thailand, for example in Nan River Basin [1], Upper Pasak River Basin [2], Bangkok and Hadyai areas [3] and in the flood risk areas [4], however, highway flooding has never been theoretically classified into various degrees of hazard. Weighting factors for each cause of the highway flooding are essential information for classifying the degree of highway flood hazard [5] which is a major step for highway flood hazard mapping and for enabling flood monitoring and warning in hazardous areas.

2 Objectives

The aim of this study is to investigate the causes of highway flood, then, to apply the multi criteria analysis (MCA) based on the analytic hierarchy process (AHP) to determine the weighting factors of importance for each cause of the highway flood for the whole Kingdom of Thailand.

3 Methodology and theoretical considerations

3.1 Methodology

The degree of flood hazard varies with circumstances across the full range of floods. In this research, the causes of highway flood were investigated using relevant historical records, namely, historical highway floods, hydrological time series and basin topographical characteristics. Then the weighting factors for each cause of highway flooding were assessed using MCA based on AHP [6, 7])

3.2 Theoretical considerations

Weighting factors are the estimated values indicating the relative importance or impact of each item in a group as compared to the other items in the group. The purpose of assigning the weighting factors is to help us define the level of importance or impact of criteria. This process necessarily assigns numerical values to judgments. These judgments should not be arbitrary or subjective, but should reflect expert views and should be supported by objective information. To achieve the meaningful results which ones can rely on, the AHP, which was developed by Thomas L. Saaty in 1970, was used in this study. This is because AHP is a structured technique for organizing and analyzing complex decisions based on mathematics and psychology. The AHP application involves the mathematical synthesis of numerous judgments about the decision problem at hand. The procedure can be subdivided into 4 steps, namely problem decomposition, establishing priorities, consistency check and final conclusion. The details of each step are as follows:

Step 1 Problem decomposition: The first step in the analytic hierarchy process is to decompose the decision problem into a hierarchy of more easily comprehended sub-problems. The elements of the hierarchy can relate to any aspect of the decision problem either tangible or intangible. It may contain the decision goal, the alternatives for reaching it, the factors influencing or the criteria for evaluating the alternatives.

Step 2 Establishing priorities: Once the hierarchy has been constructed, the priorities, the importance or the impacts among the elements of the hierarchy will be established by decision maker(s) or expert(s) by making a series of judgments based on pair-wise comparisons of the elements. The criteria are pair-wise compared against the judgment for importance. The comparisons are mathematically processed, and the priorities are derived. The AHP converts these evaluations to numerical values that can be processed and compared over the entire range of the problem. The priorities are absolute numbers within a range of numbers from 'least effect than', to 1 for 'equal', and to the 'absolutely more important than' covering the entire spectrum of the comparison. Table 1 illustrates the sample of the priority matrix of a 3 criterion problem.

Table 1: Sample of the priority matrix of a 3 criterion problem.

Criteria	1	2	3
1	1	$1/X_{21}$	$1/X_{31}$
2	X_{21}	1	X_{23}
3	X_{31}	$1/X_{23}$	1

Remarks: X_{ij} is the relative priority of criterion i in comparison with criterion j.

Step 3 Consistency check: The consistency of the judgments is checked by performing the calculations to find the maximum eigenvalues, consistency index *CI*, consistency ratio *CR* and normalized values for each criterion. If the maximum eigenvalues, *CI* and *CR* are satisfactory, then a decision is taken based on the normalized values; else, more information has to be supplied to the decision maker(s) or the expert(s), then the procedure is repeated till these values lie within an acceptable range.

The consistency index *CI* can be computed using

$$CI = \frac{\lambda - n}{n-1} \tag{1}$$

where

$$\lambda = \sum_{i=1}^{n}\left[\left(\sum_{j=1}^{n} X_{ij}\right) w_i\right], \tag{2}$$

n is the number of criteria for comparison, X_{ij} the relative priorities (pair-wise comparison between criteria i and j) and w_i the normalized weight of criterion i (see the details on how to compute this value in section “Results and discussion”).

The consistency ratio (*CR*) can be computed from

$$CR = \frac{CI}{RCI}.$$

where *RCI* is the random consistency index. This varies according to the number of criteria for comparison (n) as shown in table 2. Then *CR* is compared with the acceptable range as follows:

$$CR \leq 0.10 \qquad \text{for } n > 5$$
$$CR \leq 0.09 \qquad \text{for } n = 4$$
$$CR \leq 0.05 \qquad \text{for } n = 3$$

If the computed *CR* is higher than the acceptable range, more information is supplied to the decision maker(s) or the expert(s) and the procedure is repeated till the value lies within the acceptable range.

Table 2: Random consistency index (*RCI*).

n	1	2	3	4	5	6	7	8
RCI	0.00	0.00	0.58	0.90	1.12	1.24	1.32	1.41
n	9	10	11	12	13	14	15	
RCI	1.45	1.49	1.51	1.54	1.56	1.57	1.59	

Remarks: n is the number of criteria for comparison.

Step 4 Final conclusion: The AHP final process will serve as a tool to reach the meaningful and objective numerical value of each criterion. The final step involves the important task to determine the weighting factors indicating the relative importance or impact of each criterion. In the final step, the numerical weighting factors are calculated using the average values of normalized weights of each criterion obtained from Step 3 of the process by all decision maker(s) or expert(s). It can be determined as follows:

$$W_i = \frac{1}{m}\sum_{j=1}^{m} w_{ij} \text{ for } i=1, 2, ..., n \tag{3}$$

where W_i is the weighting factor for criterion i, w_{ij} the normalized weight of criterion i by decision maker j and m the number of decision maker(s) or expert(s).

4 Study area

Thailand covers an area of 513,115 km^2. The country is bordered by Malaysia in the south, the Union of Myanmar in the west and northwest, the Loa People's Democratic Republic to the northeast and Cambodia to the southeast. As reported in December 2012, the estimated population was about 64.6 million with a growth rate of 0.32 percent.

The country can be subdivided into five main geographical regions, namely, the north, the central plains, the northeast, the east and the south. The northern region is mainly mountainous. It is the origin of four major rivers (Ping, Wang, Yom, and Nan) which converge to become the Chao Phraya River in the central plains. The northeast region occupies one-third of the country. The area is a high plateau at 100 to 200 m elevations. The eastern Thailand is bound by the Gulf of Thailand in the south and Cambodia in the east. The southern region of the country is located on the peninsula between the Andaman Sea of the Indian Ocean in the west and the South China Sea of the Pacific Ocean in the east.

The location of Thailand, the study area is shown in fig. 1. The average annual temperature varies in the range from 23.3 to 28.4 °C. The average annual rainfall for the whole country is about 1,425 mm. The total annual rainfall of all river basins is about 800,000 million m^3 of which 75% is lost through evaporation and evapotranspiration. The remaining 25% (200,000 million m^3) is stored in streams, rivers and reservoirs.

According to the information from the Department of Highways, as of September 2013, the total length of highways in the country was 51,610.888 km. It consists of 15,720.566 km in the northern region, 15,116.624 km in the northeastern region, 10,881.481 km in the central region and 9,892.217 km in the southern region. Fig. 2 shows the map of the highway network in Thailand.

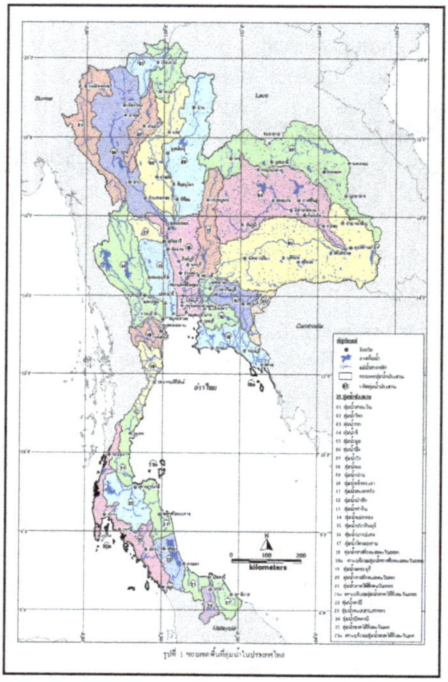

Figure 1: Location of study area [8].

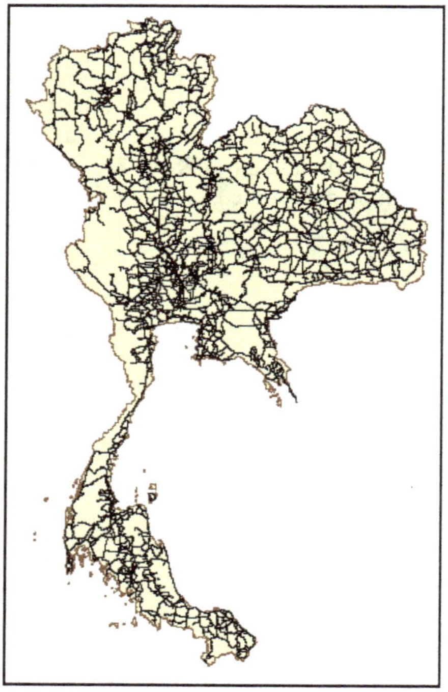

Figure 2: Highway network in Thailand [5].

5 Data collection

Historic records of highway flood, hydrological time series and land topography were collected from the official database of the Department of Highways (DOH), the Thai Meteorological Department (TMD), the Royal Irrigation Department (RID), the Water Resources Department (WRD), the Department of Public Works and the Town and Country Planning (DPT) and the Department of Disaster Prevention and Mitigation (DDPM). In correspondence with the task, the following information was collected:

1) Historic records of highway floods in the whole kingdom of Thailand during 1995–2012.
2) Daily and monthly climate and rainfall time series for the period of 61 years from 1952 to 2012 from 158 synoptic stations spatially distributed in the whole kingdom of Thailand,
3) Momentary peak discharge and gauge height from 229 synoptic stations spatially distributed within 25 major river basins in the whole kingdom of the country as shown the locations of all stations in fig. 3.

Figure 3: Location of hydro-meteorological station.

6 Results and discussions

6.1 Highway flood causes

One of the objectives of this study is to investigate the causes of highway flood. Upon learning of the highway flood events within the country by taking into account the available resources and significance of the events, namely historical highway floods, hydrological time series, topography and stream conveyance capacities, it was found that highway flooding resulted from a combination of factors. The major causes are highway characteristics, hydrological characteristics and topographical characteristics. The highway characteristics play important roles in flooding. This is simply due to the drains or the culverts being blocked or insufficient for large amounts of floodwater or flash storms. Drains can become blocked as a result of leaves or other debris covering the drain inlet, pipes blocked with debris, silt or tree roots and deteriorated or cracked pipe drains. This causes this type of flood on highways on a frequent basis. Based on the data collected, it was found that out of 17 year historic records of highway flood in the whole kingdom of the country during 1995–2012, the maximum number of flood years in a highway is 14.

The hydrological characteristics are another major highway flood cause. This is due to high river water levels preventing surface water draining from the highway, or the overbank flow due to the floodwater exceeding the conveyance capacities of the channels. In this research, the momentary peak discharge for the past 61 years during 1952–2012 from 229 synoptic stations spatially distributed within 25 major river basins in the whole kingdom of the country were used to compare with the conveyance capacity of the streams near to the highways. The stream conveyance capacity was evaluated using the Manning formula as follows:

$$Q = \frac{1}{n} R^{2/3} S^{1/2} A \tag{4}$$

where Q is the discharge (cms), n the Manning's coefficient which is defined based on the channel characteristics incorporated in previous studies by various agencies, A the cross sectional area (m^2), P the wetted perimeter (m), S the river bed slope and R the hydraulic radius (m) given by

$$R = \frac{A}{P}$$

Overlaying the historical flooding of highways near the streams with the channel conveyance capacities, it was found that overbank flooding occurred when flow events exceed the bank full capacity of the channel and therefore spill over into the highways. Hence, less stream conveyance capacity than the momentary peak discharge implies higher chance of overbank flood.

According to the list of flood risk areas collected from the Department of Disaster Prevention and Mitigation, the basin's topography is another significant factor contributing to the highway flooding. Highway on flood-prone areas for example on floodplain, low-lying or flat areas would naturally be affected by river overbank flooding and high tides or storms in coastal areas. In addition, the highways on steep landscapes are vulnerable to flash flood.

6.2 Weighting factors of highway flood causes

The results of MCA based on the AHP to determine the weighting factors of importance for each cause of the highway flood for the whole Kingdom of Thailand are presented in the following sections.

6.2.1 Problem decomposition

The first step of the AHP is to decompose the decision problem into a hierarchy of more easily comprehended sub-problems. In this study, the problem was explored as the criteria or causes/factors influencing the highway flood, namely highway characteristics, hydrological characteristics and land topography. The importance of the inundated area, depth and time of highway flooding is considered in this study to be equal. This is due to the reason that the flood at any depth and any time span jeopardizes human lives and causes losses. In order to avoid the potential risk to life, their importance is therefore considered to be equal. A schematic of AHP hierarchy for weighting factors determination of the highway flood causes is shown in fig. 4.

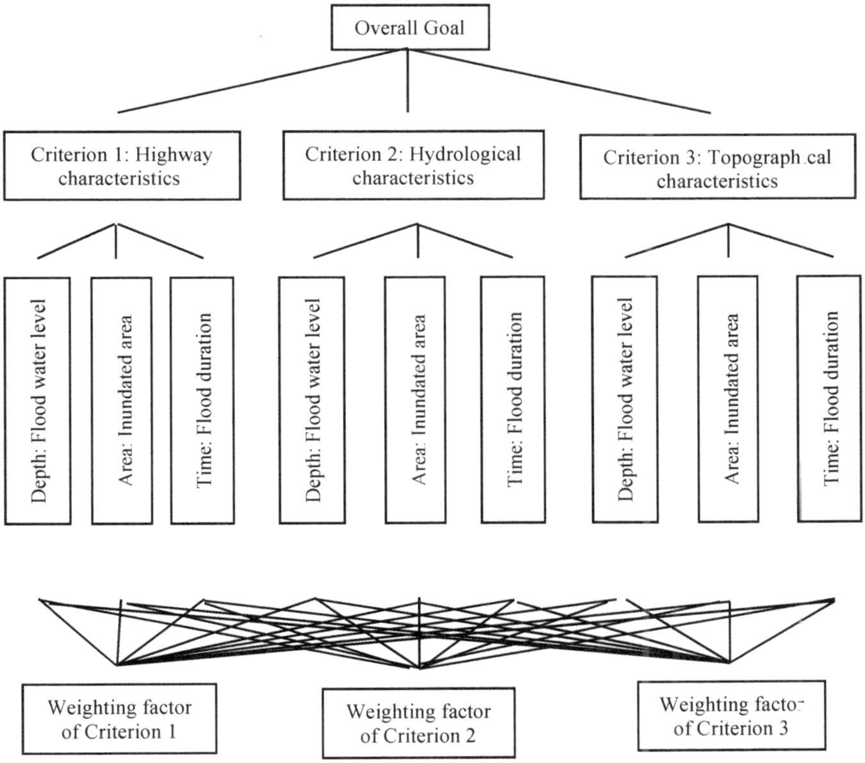

Figure 4: Schematic of AHP hierarchy to determine the weighting factors influencing the highway flood.

6.2.2 Establish priorities

Priorities among the elements of the hierarchy were established by making a series of judgments based on pair-wise comparisons of the elements. This implies what influence one impact has on the cause of highway flooding in comparison with another one in the hierarchy. In this study, the priorities are absolute numbers between zero and nine. The scale ranges from 0 for 'least effect than', to 1 for 'equal', and to 9 for 'absolutely more important than' covering the entire spectrum of the comparison as shown in table 3.

To clearly present how to proceed with the AHP, the weight given by one of our researchers to the criteria influencing the causes of highway flood is illustrated in table 4. The authors considered that the highway characteristics and the hydrological characteristics play equal role in highway flooding (level of importance is 1). But the highway characteristics have higher influence on highway flooding than the topographical characteristics, but not by much (level of importance is 2). Also, the authors decided that the hydrological characteristics have a slightly higher influence on highway flooding than the topographical characteristics (level of importance is 2).

Table 3: Level of importance of each criterion.

Level of importance	Meaning of Importance	Description
1	Equal	Both criteria play equal influence to the cause of highway flood.
3	Moderate	One criterion plays moderate higher influence to the cause of highway flood than another one
5	Very important	One criterion plays significant higher influence to the cause of highway flood than another one
7	Very, very important	One criterion obviously plays significant higher influence to the cause of highway flood than another one
9	Extremely important	One criterion plays extreme significant higher influence to the cause of highway flood than another one
2, 4, 6, 8	In between	

Table 4: Illustrated matrix of priorities of the 1st researcher.

Criteria	Highway characteristics	Hydrological characteristics	Topography characteristics
Highway Characteristics	1	1	2
Hydrological Characteristics	1	1	2
Topography Characteristics	0.5	0.5	1

6.2.3 Consistency check

The consistency of the judgments was checked by calculating the consistency index *CI* using eqn (1) with $n = 3$. To determine λ using eqn (2), the normalized weights w_i of each criterion were computed. For the highway and the hydrological characteristics, the relative priorities (pair-wise comparison between criteria *i* and *j* are listed in columns 1 and 2 of table 4. With these values, the normalized weights (w_i) are

$$\begin{bmatrix} 0.4 \\ 0.4 \\ 0.2 \end{bmatrix}.$$

For the topographical characteristics, the relative priorities are given in column 3 of table 4. With these values, the normalized weights (w_i) are equal to

$$\begin{bmatrix} 0.4 \\ 0.4 \\ 0.2 \end{bmatrix}.$$

The average normalized weights are therefore equal to

$$\begin{bmatrix} 0.4 \\ 0.4 \\ 0.2 \end{bmatrix}.$$

Substituting the above values in eqn (2) gives $\lambda = 3$ and from eqn (1), $CI = 0$.

As shown in table 2, for $n = 3$, RCI is 0.58. The consistency ratio (CR) can then be computed as $CR = 0$, which lies within the acceptable range (acceptable range is ≤ 0.05 for $n = 3$). This implies that the weight given to the criteria influencing the causes of highway flood based on AHP by the authors is consistent. It means that the authors considered that the highway and the hydrological characteristics were equally weighted at a 40% importance as the cause of highway flooding. The land topography was weighted to have a 20% influence on the cause of highway flooding.

In the context of the present study, two other groups of researchers made a series of judgments based on the pair-wise comparisons of the elements by deciding all three criteria play equal roles in highway flooding. Then, the consistency of the judgments was checked as described in earlier sections.

6.2.4 Weight of criteria influencing highway flood causes

In the final step, the numerical weighting factors given to the criteria influencing the causes of highway flooding were calculated using the average values of normalized weights of each criterion obtained from all researchers as shown in table 5. It was then concluded that based on the AHP results, the highway and the hydrological characteristics were equally weighted at a 35% importance as the cause of highway flooding. Also, the land topography was weighted as having a 30% influence on the cause of highway flooding.

Table 5: Results of weights given to the criteria influencing the highway flood.

Criteria	Researchers			Average
	1st	2nd	3rd	
Highway characteristics	0.4	0.33	0.33	0.35
Hydrological characteristics	0.4	0.33	0.33	0.35
Topographical characteristics	0.2	0.33	0.33	0.30

7 Conclusions

Flooding is one of the major natural disasters which frequently affect millions of people in Thailand. To help reducing the highway flood hazard, and preparing the information for further use to determine the highway flood hazard level, the causes of highway flooding and the weighting factors of criteria influencing highway flooding were investigated. Historic records of highway floods in the whole kingdom of Thailand during 1995–2012, historic records of a 61-year time series during 1952 to 2012 from synoptic stations spatially distributed in the whole kingdom of the country and land topographical data were used in this part of study. It has been found that the major causes of highway flooding in Thailand are the highway characteristics, the hydrological characteristics and the land topographical characteristics. According to the results using the multi-criteria analysis based on the analytic hierarchy process, the highway and the hydrological characteristics were equally weighted as having a 35% importance as the cause of highway flood. Also, the land topography was weighted as having a 30% influence on the cause of highway flooding. The weighting factors of criteria influencing highway flooding obtained in this study are applicable to set the priority for working out solutions to the problems and help reducing the flood incidents. In addition, these factors are essential information for further study to classify the degree of highway flood hazard for enabling flood monitoring and flood warning in hazardous areas. Further study on highway flood hazard mapping is recommended.

References

[1] Somchit, A., Fowze, J.S.M., Bormudoi, A., Khazaroka, A. & Samarakoon, L., Flood hazard mapping in Nan River Basin. *Proc. of the 7th Annual Mekong Flood Forum Bangkok*, Mekong River Commission Regional Flood Management and Mitigation Centre (RFMMC) Thailand: Bangkok, 2009.

[2] World Bank Group, Feasibility study of flood modelling in Upper Pasak River Basin, Petchaboon/Thailand, World Bank Group, 2011.

[3] Department of Disaster Prevention and Mitigation, Study on flood hazard map in developing countries, Pacific Consultants International, 2009.

[4] Department of Disaster Prevention and Mitigation, Early Warning system installation in flood and land slide risk areas for budget years 2008–2012, 2008.

[5] Budhakooncharoen, S., Working paper for highway flood hazard mapping, for Department of Highways, 2012.

[6] Saaty, T.L., A scaling method for priorities in hierarchical structures. *Journal of Mathematical Psychology*, **15(3)**, pp. 234-281, 1977.

[7] Saaty, T.L., *The Analytic Hierarchy Process*, McGraw Hill: New York, USA, 1980.

[8] 25 Major river basins of Thailand. http://water.rid.go.th/hyd/basin/25basin.htm.

The impact of flood on the socio-economic status of residents of Wadata and Gado-villa communities in the Makurdi metropolitan area of Benue State, Nigeria

R. G. Awopetu[1], S. O. Awopetu[2] & M. S. Awopetu[3]
[1]*Department of Psychology, Benue State University, Nigeria*
[2]*Nigeria Petroleum Development Company Ltd., Edo State, Nigeria*
[3]*Department of Civil Engineering, Ekiti State University, Nigeria*

Abstract

The present study examines the effect of flood on the socio-economic status of residents of Wadata and Gado-villa communities in the Makurdi metropolitan area of Benue State, Nigeria. A sample of 502 displaced residents consisting of males and females camped at St. Theresa Catholic and St. Catherine primary school, Makurdi, participated in the study. A questionnaire consisting of six sections measuring demographic variables and impacts of flood on socio-economic status such as agriculture, education, health, housing, water and sanitation was used to collect data. The study established that flood impacted negatively on the socio-economic well-being of residents in the two communities. It is recommended that the government should provide a low-cost housing estate for the flood victims as they have expressed willingness to relocate from the flood prone area so that their children can go back to school as soon as possible. Further allocation of land for residential building in the flood prone areas should be stopped by the government and private land owners. The river banks should be converted to recreation centres and green areas.
Keywords: flood, socio-economic status, residents, Makurdi, Nigeria.

1 Introduction

Flood is a natural disaster that has frequently resulted in direct loss of social and economic assets, physical injuries and other adverse consequences to the extent of psychological injuries. Natural disasters threaten lives thus causing a lot of anxiety and fear in an individual. As observed by Nasir *et al.* [1], the socio-economic impact of a flood includes the loss and destruction of property and life; this may last for a short or a long period. Serious floods in Nigeria are rare events and relatively "tame" when compared with those in other parts of the world. Their psychological impact on health varies among the population for reasons relating to population vulnerability and type of flood events [2]. In August 2012, water released from Lagdo Dam located in the northern province of Cameroon, which is up-stream of River Benue, lead to an increase in volume of River Benue and flooded several towns and villages along its course. These include Adamawa, Anambra, Edo, Taraba, Benue, Kogi, Bayelsa and Delta. The flooding resulted in more than 10 deaths and loss of properties worth several millions of naira and displaced about 131,011 people [3]. This is not the first time of such an occurrence, but the government of Nigeria is still yet to find a solution. Governments as well as the affected communities were taken completely by surprise at the severity of the event considering the fact that flood warnings were issued by the Nigeria Metrological Agency (NMET), National Emergency Management Agency (NEMA) and Federal Ministry of Environment to local residents but no action was taken. Areas that were never thought they could be flooded were completely submerged in water for days.

The term socio-economic impacts refers to all changes in the way people live, work, relate, organise and how these interactions have direct influence on means of livelihood, the purchasing and production power, mass migration and agriculture. More specifically, social impacts concern poverty, loss of life, health effects, loss of community cohesion, loss of time, changing attitudes, impoverished neighbourhood etc, which are difficult to quantify in monetary terms [4]. On the other hand, the economic impacts include disruptions of clean water and electricity supply, transport, communication, education, health care services, reduction in purchasing power and loss of land value in the flood plains which can lead to increased vulnerabilities of communities in the living area. The additional cost of rehabilitation, relocation of flood victims and removal of property from flood-affected areas can divert the capital required for maintaining production.

Many people experienced a number of associated problems which contributed to the levels of stress suffered, such as problems with personal relationships (at home and at work), problems with employment as well as lack of understanding and sympathy from authorities and society in general. People in employment found it particularly difficult to cope with all the arrangements that had to be made for getting the house back in order [5].

Studies on the socio-economic impacts of floods, like the wide impact-survey in Scotland [6] and the survey on flood experience in Belgium [7], have shown intangible impacts experienced by flood victims as being even more severe than tangible impacts. This confirms the high cost of relief and recovery which may adversely impact investment in infrastructure and other development activities in the area and in certain cases may cripple the frail economy of the communities. Recurrent flooding in a region or state/nation may discourage long- term investments by the government and private sector alike. Lack of livelihoods, combined with migration of skilled labour and inflation may have a negative impact on a nation's economic growth. Loss of resources can as well lead to high costs of goods and services, delaying its development programmes.

The social economic impacts are experienced by different people in different ways. Against this background, therefore, this paper aims at assessing the impact of flood on the socio-economic status of residents in two communities affected by the 2012 Makurdi flood, in Benue State, Nigeria.

2 The study area

Makurdi Metropolitan area is the capital city of Benue State and is located in the middle belt region (north-central) of Nigeria (6°45′–8°15′E, 7°30′–9°45′N). It has a population of 297,398 spread out over an area of 41,035 km^2 making it the most densely populated local government area in Benue State, with a mean of 257 people per km^2. It has a diverse cultural make-up, among which the following ethnic groups are prominent: Tivs, Idomas, Igedes, Hausas, Yorubas and Ibos [8].

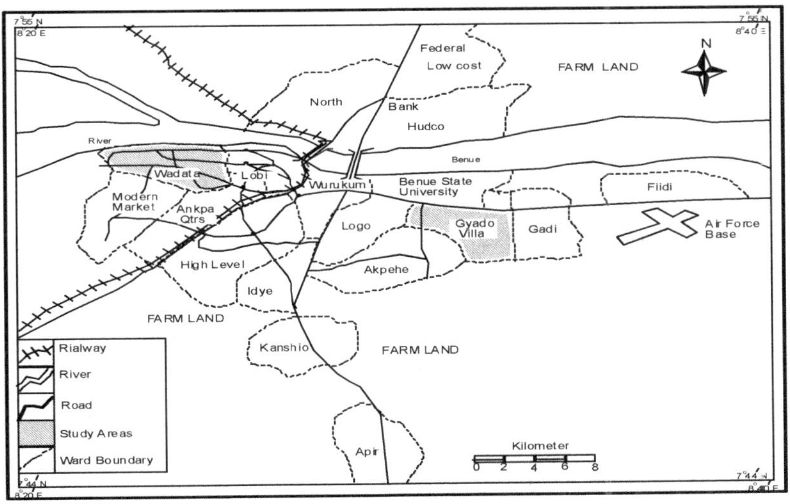

Figure 1: Map of the study area.

3 Method

The study which employed a quantitative approach used a sample consisting of victims of 2012 Makurdi flood. The victims were camped at St. Catherine and St. Theresa primary schools. A sample size of 502 victims participated in the assessment survey. A purposive sampling was adopted; this is because, according to Strydom *et al.* [9], purposive sampling is entirely based on the judgment of the researcher, in that a sample is composed of elements that contain the most characteristics, representative or typical attributes of the population.

3.1 Instruments

The instrument for the study was a household questionnaire. The questionnaire used to collect data consisted of six sections measuring demographic variables and impacts of flood on socio-economic status such as agriculture, education, health, housing, water and sanitation. Section A measures demographic variables, section B measures flood impact on agriculture, section C measures flood impact on education, section D measures flood impact on health, section E measures flood impact on housing and section F measures flood impact on water and sanitation.

3.2 Procedure

A copy of each questionnaire was given to the respondents at the relief camps. Five hundred and fifty questionnaires were administered to the respondents, by the time of retrieval; only five hundred and two were retrieved due to inappropriate fillings of some of the questionnaires.

3.3 Data analysis

Simple percentages and frequency were used in analysing the data.

4 Results and discussion

Often mostly affected by natural disaster such as flood are women and children. More than half of the respondents interviewed (52.2%) were female and more than 8 out of every 10 respondents interviewed are married (table1). Health and sanitation was a big issue. The two camps set up in a government own schools had no bathing facility. Whereas the men and young children can take baths outside on the school lawn, women do not have that option. Many people did not have a chance to pick up their belongings when the floods hit their houses so they have no change of clothes. Many are wearing what they left home in and without being able to wash and women's hygiene in particular has deteriorated. The situation is even worse for menstruating and pregnant women.

Also from table 1, it can be seen that the respondents age was categorized into five different age groups and the greater proportion of respondents falls between the age group 26–30 (30.3%) followed by age group 41 and above with 24.7%, age group 31–35 accounted for 18.9% of the respondents, age group 18–25 accounted for 13.1% of the respondents and age group 36–40 is the closest age group to the least age group with the total of 12.9%. From the data above, it is crystal clear that adults in their productive age (26–40 years), which form 62.1% of the respondents, are affected by the flood. This will impact negatively on the economy of the study area most especially when a large percentage of the respondents are self employed (fig. 2).

Table 1: Percentage distribution of respondent's socio-demographic characteristics.

Variables	Frequency	Percentage
Respondent's marital status		
Married	404	80.5
Single	66	13.1
Divorced	32	6.4
Total	502	100.0
Respondent's sex		
Male	240	47.8
Female	262	52.2
Total	502	100.0
Respondent's age group		
18–25	66	13.1
26–30	152	30.3
31–35	95	18.9
36–40	65	12.9
41 and above	124	24.7
Total	502	100.0

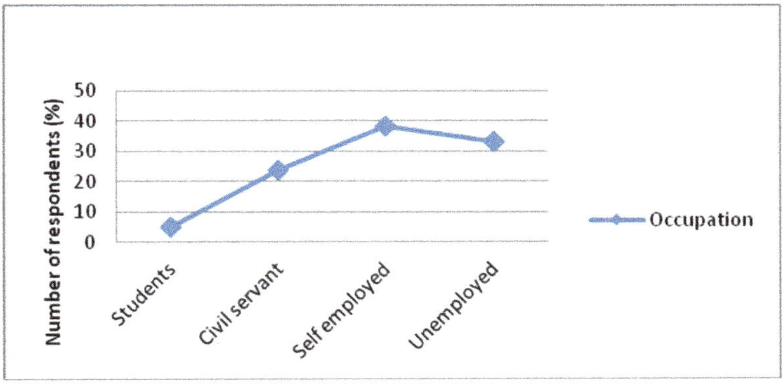

Figure 2: Occupational status of the respondents.

Furthermore, fig. 3 shows that slightly above 4 in every 10 respondents are reported to have primary education (42.4%) which implies that the majority of the respondents interviewed have primary education, followed by the respondent who reported to have secondary education (31.7%), while the respondents with tertiary education has the least percentage (25.9%). The educational statuses of the respondents are low but it cannot be generalised that the educational status of the people that are affected by the flood are generally low. This is because, flood affected people with high educational status are either gainfully employed and therefore can afford to pay for temporary accommodation or relocate to live with a friend temporarily.

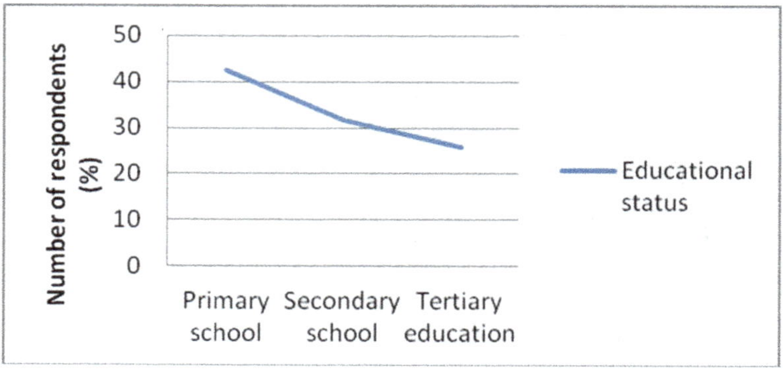

Figure 3: Educational status of the respondents.

The 2012 flood has inflicted severe direct and indirect damages on social; infrastructure and economic sectors of the Makurdi metropolis. Close to half of the respondents (48.0%) reported that their houses collapsed. About 12,000 people were displaced from their dwelling. A high number of deaths was not

Table 2: Percentage distribution of respondent's response on the impact of flood on agriculture.

Variables`	Frequency	Percentage
Household experienced crop damage as a result of flood		
Yes	342	68.1
No	160	31.9
Total	502	100.0
The main staple crop damaged as a result of flood		
Yes	360	71.7
No	142	28.3
Total	502	100.0
Household experienced loss of food stock as a result of flood		
Yes	481	95.8
No	21	4.2
Total	502	100.0

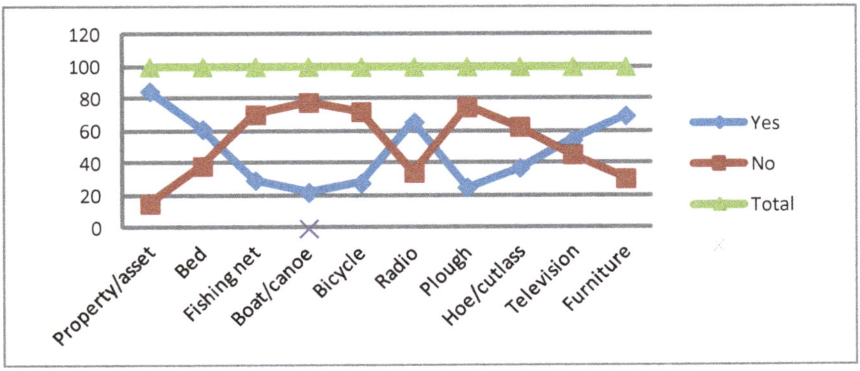

Figure 4: Percentage distribution of respondent's response on the impact of flood on housing/property/assets.

recorded because the flood gradually hit the town while people had the opportunity to remove some of their movable belongings. In spite of this, many people lost their fishing nets, canoes, furniture, electronics and plough (fig. 4). It was also revealed from table 2 that close to 7 in every 10 respondents (68.1%) reported that their household experienced crop damage as a result of flood. More than 7 in every 10 respondents (71.7%) reported that their main staple crop was damaged as a result of flood. Similarly, the majority of the respondents (95.8%) reported that their household experienced loss of food stock as a result of flood.

The impact of flood on education is described in fig. 5 where the horizontal axis categories are defined as follows:

- * Educational facilities is available in your area
- ** School infrastructures were damaged due to flood
- *** School going children in your household experience any disruption in an attendance due to the flood
- **** Road impassable is the main reason why children in your household experience any disruption in an attendance
- ***** Bridge culvert washed away or submerged is the main reason why children in your household experience any disruption in an attendance
- ****** School submerged/surrounded by water is the main reason why children in your household experience any disruption in an attendance

It can be seen from fig. 5 that educational facilities such as school buildings, learning material such as books and furniture were also affected. Worse still, the majority of the respondents (73.1%) revealed that school going children in their household experienced disruption in learning activities as well as school attendance for about three months due to the flood. More so, the table revealed that more than half of the respondents (53.6%) reported that road being impassable is the main reason why children in their household experienced disruption in an attendance Also, revealed in fig. 5 is that 44.6% of the respondents reported that bridge culvert washed away or submerged is the main reason why children in their household experience disruption in an attendance while 60.8% of the respondents reported that school submerged/surrounded by water is the main reason why children in their household experienced disruption in an attendance.

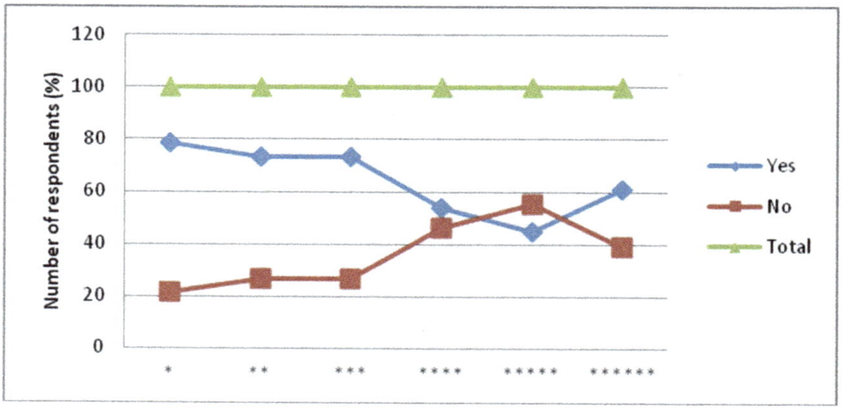

Figure 5: Percentage distribution of respondent's response on the impact of flood on education.

Health care facilities were not left out in the flood disaster. From fig. 6, it can be seen that 61.2% of the respondents reported that health facilities available in their area where damaged due to flood. More than 6 in every 10 respondents (63.1%) reported that there is disruption in access to health service as a result of the flood. Furthermore, fig. 6 revealed that more than 8 in every 10 respondents

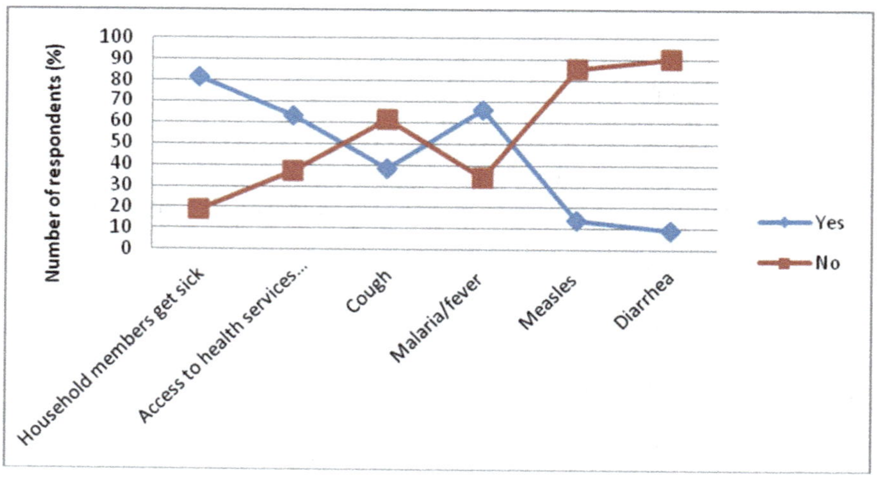

Figure 6: Percentage distribution of respondent's response on the impact of flood on health.

(81.5%) reported that their household members got sick during the flood while 9.4% of the respondent specifically reported that diarrhoea was experienced by their household member who got sick.

It can be seen from table 3 that the majority of the respondents (25.5%) reported that river water is the common source of their drinking water, 21.1% of the respondents reported that protected well is their common source of drinking water while 18.3% and 16.5% of the respondents reported that borehole and unprotected well are, respectively, their common source of their drinking water. Close to 9 in every 10 respondents revealed that their source of water are affected by the flood while 13.1% of the respondents disagreed with the fact that source of water are affected by the flood.

Table 3: Percentage distribution of respondent's response on the impact of flood on water and sanitation.

Variables	Frequency	Percentage
Common source of drinking water		
Borehole	92	18.3
Protected well	106	21.1
Unprotected well	83	16.5
River	128	25.5
Spring	21	4.2
Sachet/bottled water	72	14.3
Total	502	100.0
Source of water affected by the flood		
Yes	436	86.9
No	66	13.1
Total	502	100.0
Types of sanitary facilities available		
Vip	50	10.0
Traditional pit latrine	349	69.5
Reticulated sewerage	95	18.9
Others	8	1.6
Total	502	100.0
Our sanitary facilities affected by the flood water		
Yes	478	95.2
No	24	4.8
Total	502	100.0

Also, the majority of the respondents (69.5%) reported that traditional pit latrine is the type of sanitary facilities available in their area, 18.9% of the respondents reported that reticulated sewerage is the type of sanitary facilities available in their area while 10.0% and 1.6% of the respondents reported that VIP and others are, respectively, the type of sanitary facilities available in their area. More than 9 in every 10 respondents revealed that their sanitary facilities were affected by the flood water while 4.8% of the respondents disagreed with the fact that sanitary facilities were affected by the flood water in their area.

5 Conclusions and recommendations

Flood waters carry nutrients and sediments, which are deposited on flood plains, enriching the soil reducing or obviating the need for nourishment through artificial fertilizers. Inundated areas retain soil moisture that helps raise crops without irrigation. Rice paddies are sometimes flooded deliberately to take advantage of this natural fertilization process. A river basin is an ecological unit interconnecting spawning habitats with rearing habitats for a variety of species and other aquatic systems. Seasonal habitats on the flood plain, created by variable flow regimes, are essential for various stages of the life cycle of species. Floods provide an ecological trigger for both the spawning and migration of certain species.

The river ecosystem is a critical habitat for the biota: fish, wildlife and waterfowl. Seasonal variability and variable sediment and flow regimes help maintain comparably high levels of biodiversity in rivers and flood plains. Wetlands or swamps located in flood plains serve as natural buffer zones for excessive flood flows and play host to many birds, fish and plants. Those areas can provide a number of ecosystem services and products such as water purification, food and fibre supplies and under certain circumstances (or antecedent conditions) flood mitigation benefits. Supplementary livelihoods in the form of recreational and eco-tourism activities can be made possible by the presence of the rich river ecosystem, bestowed with abundant flora and fauna. Surface runoff and flooding can help wash down pollutants and contaminants deposited on land caused by the intensive use of pesticides and fertilizers. They also flush out accumulated organic substances brought by untreated drainage water from farmlands, stockyards, factories and domestic use and restore the ecological health of stagnant river reaches and streams by diluting them and providing clean water.

However, immediate impacts of flooding include loss of human life, damage to property, destruction of crops, loss of livestock, non-functioning of infrastructure facilities and deterioration of health condition owing to waterborne diseases. Flash floods, with little or no warning time, cause more deaths than slow-rising riverine floods. In monetary terms, the extent of damages caused by floods is on the one hand dependent on the extent, depth and duration of flooding and the velocities of flows in the flooded areas. On the other hand, it is dependent on the vulnerabilities of economic activities and communities. The huge psycho-social effects on flood victims and their families can traumatize them for long periods of time. The loss of loved ones can generate deep impacts, especially on children. Displacement from one's home, loss of property, loss of memorabilia and livelihoods, decreased levels of security in the aftermath of floods and in temporary shelters and disruption to business and social affairs can cause stress. The stress of overcoming these losses can be overwhelming and produce lasting psychological impacts.

The main conclusion is that most of the flood victims, having gone through serious hardship at their relief camps, felt that life had to continue. They were able to get back to their various houses and start life. While some were able to get back to work quickly, some are still finding it difficult to embark into normal life. It is therefore recommended that further study(s) should be conducted on the post flood assessment on the victims and traumatic experiences. Efforts should be made by government to create awareness on early warnings made by meteorological departments.

References

[1] Nasir, R., Zainah, Z. & Khairudin, R., Psychological effects on victims of the Johor flood 2006/2007, *Asian Social Science*, **8(8)**, 2012.
[2] Ahern, M., Kovats, R.S., Wilkinson, P., Few, R. & Matties, F., Global health impacts of flood: epidemiologic evidence. *Epidemiologic Reviews*, **27**, pp. 30–46, 2005.
[3] The Nation, Oct.16th, 2012
[4] Messner, F. & Green, C., Fundamental issues in the economic evaluation of flood damage. *Evaluating Flood Damages: Guidance and Recommendations on Principles and Methods*, eds. F. Messner, E. Penning-Rowsell, C. Green, V. Meyer, S. Tunstall & A. van der Veen, FLOODsite, pp. 5–21, 2007.
[5] Awopetu, R.G., Assessment of psychological impact of flooding: reports from two relief camps in Makurdi, Benue State, Nigeria. Paper presented at 2012 Annual Scientific Conference and General Meeting of the Nigerian Psychological Association (NPA), 2012.
[6] Werrity, A., Houston, D., Ball, T., Tavendale, A. & Black, A., Exploring social impacts of flood risk and flooding in Scotland. School of Social Sciences – Geography, University of Dundee, Scottish government, pp 157, 2007.
[7] Grinwis, M. & Duyck, M. Integratie van de problematic van de hoogwaterstanden en overstromingen in een socio-economische context: overstrominsellende. K.I.N.T. Brussel pp. 611, 2000.
[8] National Population Census. Federal Republic of Nigeria Official Gazette. No. 4. Lagos 19 January 2007, Vol. 94, 2006.
[9] Strydom, H., Fouche, C.B. & Delport, C.S.L., *Research at Grassroots: for the Social Sciences and Human Services Professions*, 3rd ed., Van Schaik: Pretoria, 2005.

The megacity indicator system for disaster risk management: integrated assessment of physical risks in Istanbul

E. Y. Menteşe[1], B. Ergün Konukcu[1], O. Kılıç[1] & B. Khazai[2]
[1]*Istanbul Metropolitan Municipality, Turkey*
[2]*Karlsruhe Institute of Technology, Karlsruhe, Germany*

Abstract

Decision makers need tools to understand the priorities and to set up benchmarks as well as track progress in their disaster management systems, so that they can justify decisions and investments. The megacity indicator system for disaster risk management achieves these goals by analyzing the earthquake risk under three sub-categories. Urban seismic risk, coping capacity and disaster risk management index are the main components of the system. In this study, the results for the physical risk will be presented. The physical risk is calculated for all 954 sub-districts of Istanbul with a comprehensive approach including sub-indicators such as casualties, building damages, infrastructure damages, road blockages and fire ignition probabilities. These indicators are assessed through an analytical hierarchy process and compared to each other resulting in a physical risk index based on the risk values of the sub-districts.
Keywords: disaster risk reduction, earthquake, Istanbul, indicators.

1 Introduction

With the development of a comprehensive earthquake master plan for Istanbul, the implementation of a suite of disaster risk reduction initiatives, such as microzonation studies, urban renovation and rejuvenation programs in pilot areas and the seismic strengthening of infrastructure and critical buildings, the Istanbul Metropolitan Municipality (IMM) is at a point where decisions are becoming difficult to make, as the priorities are sometimes not clear or there may be too many competing interests.

In order to overcome this obstacle, IMM has developed a Megacity Indicator System for Disaster Risk Management (MegaIST) (in collaboration with the Kandilli Observatory and Earthquake Research Institute (KOERI), the Karlsruhe Institute of Technology (KIT) and the Earthquake and Megacities Initiative (EMI) to determine thresholds for assessing the level of achievement in disaster management activities that leads to adjustment of the disaster risk reduction strategies.

The models and methodology referred to here as the "Megacity Indicators System" approach was originally developed for the Inter-American Development Bank (IDB) by the Institute of Environmental Studies (IDEA) of the National University of Colombia, Manizales (NUCM) through the IDB-IDEA Indicators Program [1]. Together with its partners at Manizales, the International Center of Numerical Methods in Engineering (CINME) of the Technical University of Catalonia and local counterparts in the Philippines, EMI undertook a preliminary application of the approach to megacities in Metro Manila [2]. In the context of EMI's Cross-Cutting Capacity Development (3cd) Program methodology and approach, Megacity Indicators System is used as an innovative risk communication tool to engage stakeholders in understanding their involvement and taking ownership of the risk factors in the city. Besides the implementation in Metro Manila, many other related applications of the model have been undertaken by the IDEA-CIMNE team, and the methodology has been tested and evaluated in other cities and sub-national regions in Latin America and Europe, including Manizales (Colombia) Quito (Ecuador), Barcelona (Spain) and Lombardy region (Italy).

The method is essentially based on an integrated disaster risk assessment approach. There have been different types of risk assessment approaches developed; by Kiremidjian *et al.* [3], Hewitt [4], Davidson and Shah [5], FEMA [6], Karaman *et al.* [7], Duzgun *et al.* [8] and many others, who all tried to quantify the level of disaster risk and vulnerability, in particular.

This study considers the integration of 3 main parameters; Urban Seismic Risk Index (*USRi*), Coping Capacity Index (*CCi*) and Disaster Risk Management Index (*DRMi*) are analyzed and assessed together to evaluate the total earthquake risk in Istanbul metropolitan area at a district level. The reason for taking "seismic" risk into account is because the project area "Istanbul" is situated mainly on an earthquake prone location. The ultimate output of the study has been the relative ranking of the districts based on their level of total earthquake risk depending on the three parameters mentioned above. However, only the physical risk assessment has been fully established so far, while work on other risks is still under way. Therefore, this paper should be approached as a theoretical background paper and a progress report.

2 Method

The Megacity Indicators System for Istanbul (MegaIST) is composed of three separate but complementary indicator sets:

- Urban Seismic Risk Index (*USRi*);

- Coping Capacity Index (*CCi*);
- Disaster Risk Management Index (*DRMi*).

The approach adopted here for developing the system originates from a holistic point of view; thus it is an integrated and comprehensive approach to guide decision-making. Evaluation of the potential physical damage (hard approach) as the result of the convolution of hazard and physical vulnerability of buildings and infrastructure is the first step of this method. Subsequently, a set of fragility conditions in society that potentially aggravate the physical effects are also considered (soft approach). Together these two steps produce a landscape of total risk for the entire metropolis as the Urban Seismic Risk Index (*USRi*) and allow decision makers to evaluate or rank sub regions (e.g. administrative districts) according to the total risk.

According to the MegaIST approach, after representing the total risk for the entire metropolitan area, the ability of a particular target audience to cope, respond and manage the effects of a disaster has to be captured through a complementary set of indicators. In defining these indicators, it is important that the needs and mandates of a particular set of decision makers is considered, otherwise the indicators would be too general to be meaningful. While the *USRi* is developed for the whole of Istanbul and can be used by any interested stakeholder, the Coping Capacity Index (*CCi*) is developed by considering the particular needs of a specific target audience – in this case the Istanbul Metropolitan Municipality. The *USRi* and *CCi* are both quantitative indicators which are based on detailed assessments of the earthquake hazard and risk in Istanbul, the vulnerability of society and its capacity to respond.

A final step in the MegaIST approach is to capture the potential for disaster risk management through a set of descriptive performance indicators and track progress (or lack of progress) on pre-defined benchmarks of corrective and prospective intervention. This concludes to the development of the Disaster Risk Management Index (*DRMi*), which, like the *CCi*, is developed to address the mandates of a target group of decision makers, however, rather than providing quantitative indicators of risk and capacity, the *DRMi* produces qualitative performance indicators. The *DRMi* are scores which are derived from a self-assessment along key functional activities policies for DRM by key stakeholders at the Istanbul Metropolitan Municipality. Fig. 1 shows how the three indicator sets of the system, complement each other but, at the same time, each indicator set is also a whole on its own.

It should be noted that in the current implementation of the MegaIST, direct and indirect economic losses are not being considered. Obviously, the economic vulnerability of a city such as Istanbul is of great importance and should be accounted for in any framework which intends to describe urban risk from a holistic perspective. However, this requires additional effort and expertise, which is currently beyond the scope of this study. In the future, a set of indicators, which reflect not only the potential for direct damage to economic assets, but also indirect or flow losses influenced by dependences on the economic production, the robustness of the supply chain and infrastructure as well as potential domino-effects, must be developed and integrated into this framework.

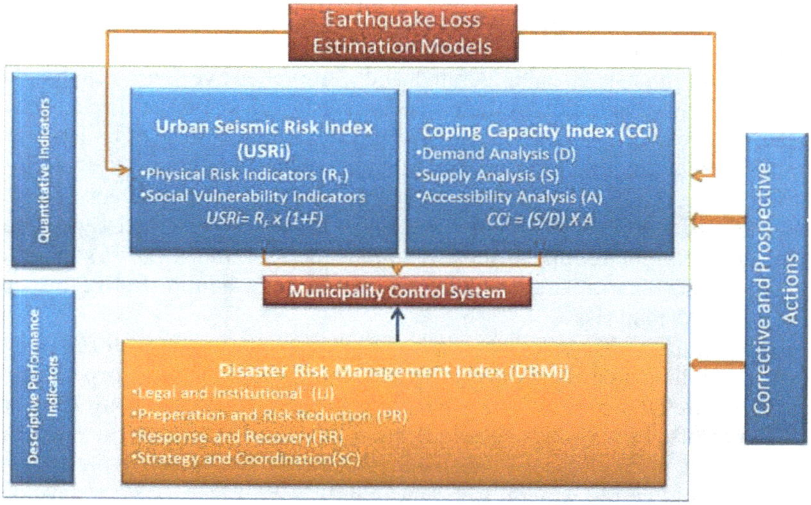

Figure 1: Conceptual model for MegaIST showing how USRi, CCi and DRMi models combine together.

2.1 Urban Seismic Risk Index

The theoretical and analytical methodological framework for the Urban Seismic Risk Index (*USRi*), based on the work of Cardona *et al.* [1] and Cardona and Birkmann [9], provide an overview of not only the expected direct damages, but also the potential for aggravating impact of the direct damages by the social fragility and lack of resilience of the different districts in Istanbul. According to this procedure, a physical risk index is obtained, for each unit of analysis, from existing loss scenarios, whereas the total risk index is obtained by factoring the risk index by an impact factor, based on variables associated with the socio-economic conditions of each unit of analysis. In the Istanbul implementation, the *USRi* combines the direct impact of expected physical damage, "physical" risk factors, and indirect "impact" factors that account for the socio-economic vulnerability within the city's population.

The potential direct impact of an earthquake, for example, is denoted as Physical Risk, R_F. The indirect effects are given by and impact factor $(1+F)$, which is based on an aggravating coefficient F. Thus the total Urban Seismic Risk Index (*USRi*) at the level of the municipality (m) is given by the following expression:

$$USRi_m = \sum_{i=1}^{m} R_{F_i}(1+F_i) \tag{1}$$

The physical risk R_F and the aggravating coefficient F, for each district (i) are in turn the weighted sums of each physical risk indicators (R_{F_j}), social fragility indicators (F_{f_k}) and social resilience indicators (F_{r_l}) in each district (i) and given by:

$$R_{F_i} = \sum_{j=1}^{n} R_{F_j} w_j \text{ and } F_i = \sum_{k=1}^{n} F_{f_k} w_{f_k} + \sum_{l=1}^{n} F_{r_l} w_{r_l} \quad (2)$$

2.1.1 Suggested indicators for physical risk

There are several aspects of the impacts induced by earthquakes and among others, physical results are the most obvious ones since they can be assessed, perceived and monitored more easily than social or economical ones [10]. Based on literature reviews and expert insights, four major sets of physical results are selected. These are casualties, building damage, infrastructure damage and fire outbreak possibility. These main themes are then subdivided into indicators that are used for physical risk analysis.

These indicators are as follows:

- Loss of lives;
- Heavy injuries;
- Building collapse;
- Building heavy damage;
- Drinking water pipeline damage;
- Sewage water pipeline damage;
- Natural gas pipeline damage;
- Road block;
- Fire outbreak.

The raw input data for physical indicators are gathered from the "*Earthquake Loss Estimation Routine-ELER*" software that is developed by Kandilli Observatory and Earthquake Risk Institute (KOERI) under the JRA-3 component of the EU FP-6 NERIES Project. The software is mainly developed for the rapid estimation of earthquake shaking and losses in the Euro-Mediterranean region [11].

The data gathered from *ELER* are all in same units corresponding to a quantity (e.g. number of collapsed buildings, number of injured persons, number of fire outbreak) except "road block" which is the percentage of the blocking of the roads in the reference grid. The software uses a grid-based system to estimate the damage which is constituted of 555 m × 418 m grids.

In order to use the data at a managerial level, the data had to be transformed into a jurisdictional base and since the smallest unit is a sub-district in Istanbul; the grids are converted to a sub-district level in GIS environment (fig. 2).

2.2 Suggested indicators for social vulnerability (S_v)

As stated above, in this study, social vulnerability is accepted as a factor that increases the impact level of physical risk. The main themes for the social vulnerability are demographic state, educational level, health state, mobility, as well as solidarity and community preparedness level of households. It must be noted that, lack or scarcity of these themes are assumed to increase the social vulnerability of the community [9]. In this study, social vulnerability indicators are determined as follows: demographic structure of households, economic structure, disability and special treatment needs, education status, access to health service, mobility and community preparedness

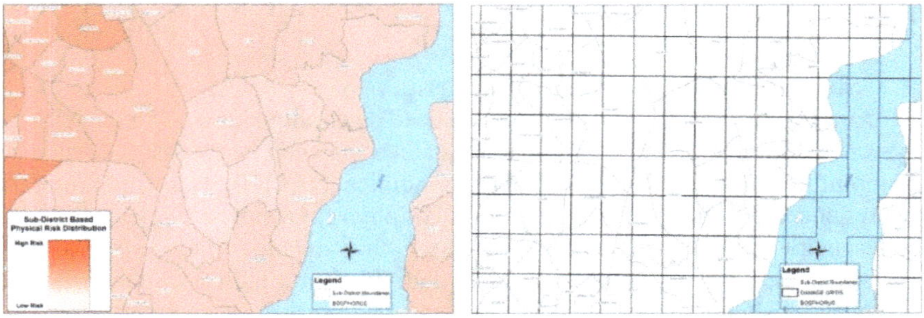

Figure 2: Conversion from grid base format to sub-district based.

The input data for social vulnerability can only be gathered via detailed surveys. This has been accomplished at pilot level yet covering 50 of the 934 sub-districts of Istanbul. Based on the official opinion of Turkish Statistical Institute, the survey requires around 35,000 household samples in order to represent the whole metropolitan area at sub-district level and household based.

2.3 Coping Capacity index (*CCi*)

The municipality's operational capacity to cope with the consequences of an earthquake constitutes the Coping Capacity indicator (*CCi*), which is quantitatively estimated for each district. The *CCi* provides stakeholders at IMM with a tool to evaluate in which districts it is lacking operational resources and where the supply in its resources may meet or exceed the demand that is derived from the risk assessment. The *CCi* and *USRi* are complementary indicator sets and can be used together to see for example which districts have the greatest total risk and also those that lack most the capacity to respond. The *CCi* and *DRMi* are also complementary indicators. Where *CCi* quantitatively describe the municipalities' operational capacity to cope and respond to a major earthquake in Istanbul in terms of its resources, the *DRMi* are performance indicators which qualitatively describe the functional capacity of the municipalities in terms of its disaster risk management policies and strategies. Another important distinction is that *DRMi* are not district-specific and are estimated for the entire city, whereas *CCi* and *USRi* are computed for each district in Istanbul.

The total coping capacity at the level of municipality (m) is the weighted sum of four types of coping capacity at IMM (j = debris removal, rescue and relief, lifeline restoration, and shelter site support) over the number of different districts in Istanbul (i), according to the following expression:

$$CCi_m = \sum_{i=1}^{m} \sum_{j=1}^{4} C_{ij} w_j \tag{3}$$

Capacity for each indicator type is defined as a simple ratio of the supply of available resources at the municipality to the demand made on these resources as well as accessibility of these resources:

$$Capacity = \frac{Supply}{Demand} \times Accessibility \quad (4)$$

In this way the results of a scenario risk assessment which are used to determine the demand on resources are factored directly into the calculation of capacity and allow IMM to leverage the results of the detailed risk study in the context of its own resource allocation and planning needs.

2.3.1 Suggested indicators for coping capacity

The aim here was to define coping capacity indicators based on the direct mandate and responsibilities of the various departments and directorates at IMM in responding and coping with a disaster in Istanbul. The disaster management center of IMM (AKOM) houses the information on where IMM's operational capacity resides and an operational flow model which identifies the roles and responsibilities of IMM's various institutions to respond and cope with the consequences of an earthquake.

According to the AKOM framework, the municipality's operational activities in responding and recovering from an earthquake constitutes the management of immediate response, relief and recovery efforts, including the removal of debris, search and rescue operations and fire extinguishing after an earthquake. It also includes the return to normalcy in the provision of fundamentally important services such as transportation, gas, water and sewage systems as well as supporting the shelter sites with basic services. Thus, the following four indicators are defined as the coping capacity indicators of IMM: debris removal, rescue and relief capacity (search and rescue, firefighting, burial capacity), lifeline restoration (natural gas, sewage, drinking water) and shelter site support (food distribution, water distribution, bread distribution) capacity.

2.4 Disaster Risk Management Index

The Disaster Risk Management index (*DRMi*) is used as a "control system" in the MegaIST framework, for the Municipality to measure performance and effectiveness of different operational and organizational policies and their effect on the total Urban Seismic Risk in the different districts of Istanbul. *DRMi* permits a systematic and quantitative benchmarking of different functional capacities and policies at the municipality during different periods. The *DRMi* are "macro" indicators for measuring progress on disaster risk reduction management for the Municipality of Istanbul. They are presented in four broad indicator groups: legal and institutional requirements ($DRMi_{LI}$), risk reduction implementation and preparedness activities ($DRMi_{PR}$), readiness to respond and recover ($DRMi_{RR}$) and strategy and coordination ($DRMi_{SC}$):

$$DRMi = DRMi_{LI} w_{LI} + DRMi_{PR} w_{PR} + DRMi_{RR} w_{RR} + DRMi_{SC} w_{SC} \quad (5)$$

3 Implementation approach

Each of the indicator sets depicted in MegaIST are derived from a set of quantitative or descriptive indicators, which must be assigned importance weights. In the case of the quantitative indicators they first have to be transformed between the values of 0 and 1. Thus the procedure for implementing the MegaIST methodology involves the following four steps: collecting and analyzing data for selected indicators; developing transformation functions to normalize indicators; implementation of indicators and their transformation functions with a set of "borrowed" weights in the Megacity Indicators System and generating the index for the selected project area.

To aid in the implementation process, an interactive software tool, Logical Decisions for Windows (LDW) is used to evaluate and qualify indicators and examine their effects upon the total output. The implementation in Manila has been tested and validated with the use of the software, which was found to be a flexible and practical analytical platform for duplicating the proposed methodology. The software supports performing sensitivity analyses, which can be used to interactively demonstrate variability of the results to different indicators. It also enables an evaluation of the stability of the methodology to variability of the input data. The advantage of using such a tool that integrates the models is that during a periodic evaluation, other indicators that may have previously not been available or simply overlooked can be integrated into the framework to obtain a new evaluation of risk and Disaster Risk Management (DRM) practices in Istanbul.

4 Integrated physical risk analysis

The hazard study that is based on this analysis results from a scenario earthquake on the main Marmara Fault (fig. 3) with Mw=7.25 and $\varepsilon = 1.4$ with probability of exceedance 10% in 50 years (475 year return period).

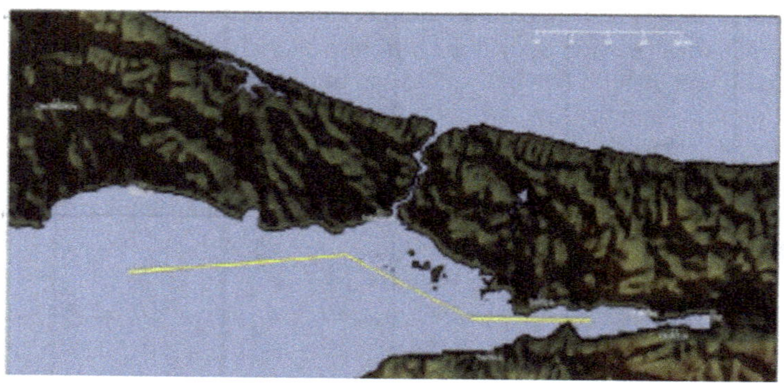

Figure 3: Approximate fault segment in Marmara Sea.

As mentioned in detail in the methodology, the dataset of ELER is grid based [11], and it does not represent full efficiency with respect to its use in decision making procedures. Thus the data had to be converted to a sub-district base.

According to ELER results, the most probable earthquake scenario in Istanbul will result in approximately 5,000 completely damaged buildings, 20,000 extensively damaged buildings, 16,000 deaths and 16,000 serious injuries, 456 damages on water pipeline network damage, 1478 damages in sewage water pipeline network and 644 damages in natural gas pipelines [12]. In addition, 423 flammable buildings are estimated to be damaged which are assumed as the main sources of fire outbreaks following an earthquake.

After the grid based ELER results (number of completely and extensively damaged buildings, number of deaths and seriously injured people, percentage of road blockage, number of damages on drinking water, waste water and natural gas pipeline network and damage to flammable buildings and wooden buildings for analyzing fire outbreak possibility) were converted to sub-district base, they were embedded in the LDW software database.

Thus, each sub-district is represented with their damage values in the database. Then these values are normalized, that is, transformed into common units varying between 0 and 1 where 1 is the highest risk and 0 is the lowest. A screenshot from LDW software can be seen in fig. 4 where the depiction of relation between casualty and risk is represented.

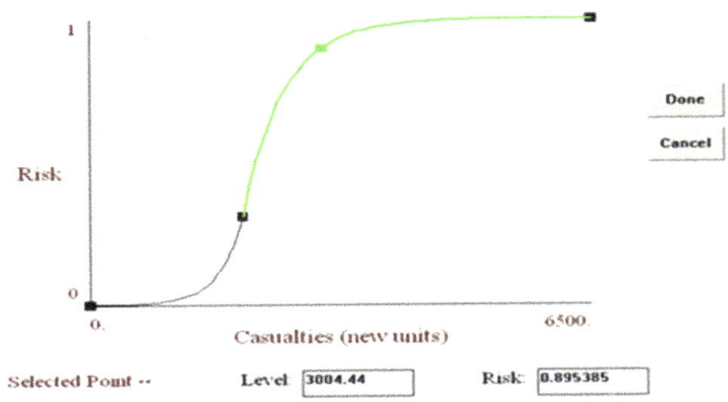

Figure 4: Normalization process.

Thereafter, the damage values are weighted based on expert and stakeholder opinions using the Analytical Hierarchy Process in the LDW software (fig. 5). It must be noted that weightings were done with the perception that priority must be given to the indicator that is the most important and precautions must be taken in this regard. It is seen that the main opinion of the experts is that building damage and its inevitable output loss of lives or injuries must be handled much more rapidly than any other indicator. These indicators consist of almost 75% of the whole importance weights.

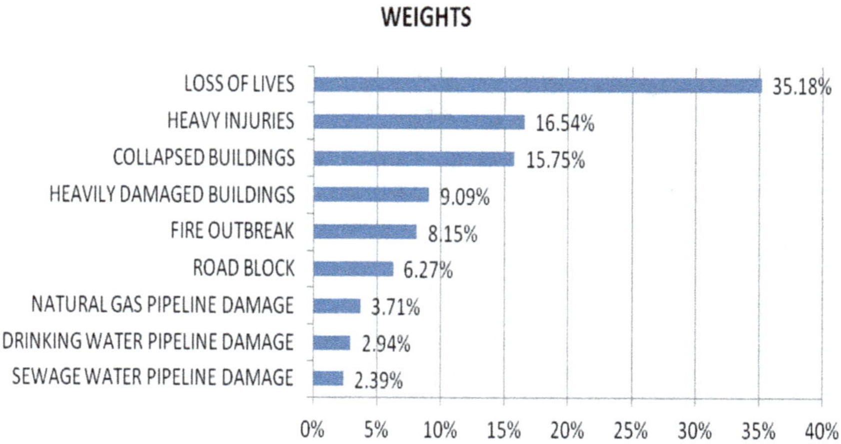

Figure 5: Percentage weights of the physical indicators.

After the weightings are done, each sub-district gets a risk value varying between 0 and 1. This data basically depends on the damages but the latter are normalized to be indexed and weighted based on their importance. As a result, the total physical risk map of Istanbul is generated as seen in fig. 6.

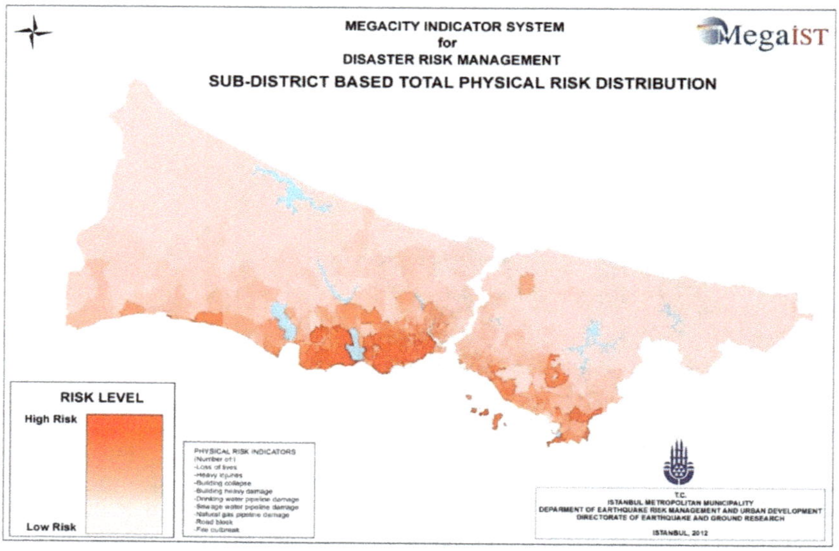

Figure 6: Sub-district based physical risk level distribution in Istanbul.

5 Results and discussion

According to the results, it is observed that, especially the southern part of the European side of Istanbul is prone to risk more than other parts of the whole metropolitan area. It is because this area has the maximum numbers of casualty and building damages that also have the highest weights.

The map also shows that the urban structure of Istanbul lacks earthquake safety, since rural areas are mostly performing well with regard to resiliency; but urbanized areas have higher risk values. This is definitely a result of uncontrolled and unauthorized urbanization process in the last 50 years. The results also show correlation with the JICA study [13] although having different numbers, the damage pattern shows criticality in the same region.

The significance and uniqueness of this analysis is that, it is not solely a damage estimation approach; it is a model that combines various types of risks and generates an integrated and holistic output that enlightens the decision makers on where they should start to reduce risks. To be able to prioritize the disaster risk reduction actions is crucial in a risk prone city such as Istanbul that can save thousands of lives. Another important part of the study is that it can give input on how the stakeholders and experts react to possible risks; how they weight the importance of these indicators.

This analysis has been conducted as a part of MegaIST project which is explained in detail in the first sections. In the next two years, the aim is to develop and complete the other parts of the project and finalize the overall index constituted of not only physical, but also social, institutional and organizational aspects of disasters.

Acknowledgements

The authors of this paper would like to thank Mr. Fouad Bendimerad (President of Earthquake and Megacities Initiative) who motivated and inspired us in the first place. In addition, we would like to thank Prof. Mustafa Erdik, Head of Kandilli Observatory and Earthquake Research Institute (KOERI) for his technical assistance and invaluable consultancy, Mrs. Bilgen Sungay again from KOERI for her active participation in our discussions and Istanbul Metropolitan Municipality for providing us technical and administrative conditions to carry out this study.

References

[1] Cardona, O.D., Hurtado, J.E., Duque, G., Moreno, A., Chardon, A.C. & Velásquez, L.S., System of Indicators for Disaster Risk Management, Main Technical Report, Manizales, 2005.

[2] Earthquake and Megacities Initiative, Application of Indicators in Urban and Megacities Disaster Management, A Case Study for Metro Manila. Progress Report, Manila, 2006.

[3] Kiremidjian, A., Moore, J., Fan, Y.Y., Yazlalı, O., Basoz, M. & Williams, M., Seismic risk assessment of transportation network systems.*Journal of Earthquake Engineering*, **11**, pp. 371–382, 2007.

[4] Hewitt, K., *Regions of Risk: A Geographical Introduction to Disasters*. Longman: Essex, 1997.

[5] Davidson, R. & Shah, H.C., A multidisciplinary urban earthquake disaster risk index. *Earthquake Spectra,* **13(2)**, pp. 211–223, 1997.

[6] Federal Earthquake Management Association (FEMA), HAZUS: user's manual and technical manuals. vol 1–4., Washington, DC, 1999.

[7] Karaman, H., Sahin, M. & Elnashai, A.S., Earthquake loss assessment features of Maeviz-Istanbul (Hazturk). *Journal of Earthquake Engineering*, **12**, pp. 175–186, 2008.

[8] Duzgun, H.S.B., Yucemen, M.S., Kalaycioglu, H.S., Celik, K., Kemec, S., Ertugay. K. & Deniz A., An integrated earthquake vulnerability assessment framework for urban areas. *Natural Hazards*, **59(2)**, pp. 917–947, 2011.

[9] Cardona, O.D. & Birkmann, J., (eds). A system of indicators for disaster risk management in the Americas.*Measuring Vulnerability to Hazards of Natural Origin: Towards Disaster Resilient Societies,* United Nations University Press: Tokyo, 2006.

[10] Carreño, M.L, Cardona, O.D. & Barbat, A.H., A disaster risk management performance index. *Natural Hazards*, **41(1)**, pp. 1–20, 2007.

[11] Kandilli Observatory and Earthquake Research Institute (KOERI), Earthquake Loss Estimation Routine (ELER) Damage Loss Estimation Software, User Manual. http://www.koeri.boun.edu.tr/depremmuh/eski/ELER/ELER_v3_Manual.pdf

[12] Kandilli Observatory and Earthquake Research Institute (KOERI), Updating Estimations of the Probable Earthquake Loss Estimations in Istanbul, Project Report, Istanbul, 2009.

[13] Japan International Cooperation Agency, Disaster Prevention and Mitigation Basic Plan in Istanbul including Seismic Microzonation, Final Report, Istanbul, 2002.

Seismic vulnerability assessment of health facilities in Mexico City, Mexico

S. Morán-Rodríguez[1] & D. A. Novelo-Casanova[2]
[1]*Instituto de Ecología A.C. de México, Veracruz, Mexico*
[2]*Universidad Nacional Autónoma de México, Mexico*

Abstract

A model was developed to estimate the seismic vulnerability of health facilities; it consists of three phases: (1) design of a theoretical framework (TF) considering structural, non-structural, functional, and social-administrative elements; (2) identification of the vulnerability conditions of the analyzed structure using the TF; (3) estimation of the level of vulnerability to seismic hazards of the analyzed health facility by comparing the measured vulnerability of the installation with the optimal theoretical indicators. This methodology was applied to three hospitals in Mexico City: the Cardiology Hospital of the National Medical Center "Siglo XXI" (CaH), the Children's Hospital "Dr. Federico Gómez" (ChH) and the "Hospital de Jesus" (JeH). The results indicate that the JeH is the most vulnerable of the three cases of study with unacceptable level of vulnerability in the social-administrative element. In the functional and the non-structural elements, this facility has very high level of vulnerability. Besides, the JeH has very high structural vulnerability level in one of its buildings. The ChH's social-administrative element was classified with moderate level of vulnerability. However, the vulnerability of its functional and non-structural elements is high. In the case of the structural element of this facility, two of its buildings have very high vulnerability level. The CaH was classified with very high level of vulnerability in the socio-administrative element. However, the functional and non-structural elements are high. The vulnerability of the structural element of three buildings located in this hospital is moderate and only one building was identified with low level of vulnerability. In general, the main causes of the unacceptable, very-high and high levels of vulnerability measured in the three hospitals are due to: (1) the architectural shape of the buildings, type of construction and the poor technology to withstand seismic shaking; (2) the lack of a proper operational emergency plan; (3) problems related to roofs' overweight and connecting bridges among the different installations.
Keywords: health facilities, vulnerability, Mexico City, seismic hazard.

1 Introduction

Although health facilities are essential infrastructure during disasters and emergencies in any urban area, they also are usually highly vulnerable installations in the case of the occurrence of a local large earthquake. Thus, it is important to maintain the operability of a hospital after the impact of an earthquake to satisfy the need for medical care of the affected population [1].

Damage in hospitals can cause devastating secondary disasters. On December 2004, the Indian Ocean tsunami disaster affected the entire national healthcare system and millions of beneficiaries of the Aceh province in Indonesia. It damaged 61% of health facilities and killed approximately 7% of its health workers [2]. As a consequence, Aceh's public health system recovery required intensive investment. For these reasons, it is important to carefully design the structure as well as the geological and geographical location of hospitals to minimize their exposure to seismic hazards [3, 4]. Also, it is necessary to implement non-structural mitigation measures and legislate for building regulations of countries located in areas highly exposed to seismic hazards [5, 6].

The most recent devastating earthquake in Mexico City occurred on 19 September 1985 (magnitude 8.1) [7]. During this earthquake, five main health facilities suffered major damage. At least 11 hospitals had to be evacuated and in total, 856 people lost their lives in two collapsed hospitals [8, 9]. Also, 13 hospitals of six or more floors were partially or totally destroyed. One out of every four hospitals' beds in the city was lost. More than 900 patients, physicians, nurses and paramedical workers died in the initial shock [10]. Moran-Rodríguez and Novelo-Casanova [11] determined that about 70% of the most important health's infrastructure in Mexico City has a level of exposure to seismic hazards between moderate and high. Although the other 30% has low level of exposure, its level of damage during a large earthquake will depend on its vulnerability conditions. These authors identified that more than 50% of the main and reference hospitals in Mexico City are highly exposed to seismic hazards.

At present, there are different methodologies to estimate seismic hazards. Cardona and Hurtado [12] developed a holistic approach to estimate seismic risk incorporating geological and structural elements as well as social, political and cultural aspects, even the responsiveness and recovery system after the occurrence of an earthquake. The method uses mathematical tools to estimate the seismic hazard in cities or urban areas determining a Risk Management Indicator (RMI). This methodology allows measuring key factors of countries' vulnerability to natural hazards and the level of performance of the local disaster risk management [13].

The community vulnerability assessment tool (CVAT) is a risk and vulnerability assessment methodology designed by the National Oceanic and Atmospheric Administration's Coastal Services Center (NOAA) to assist emergency managers and planners in their efforts to reduce vulnerability through hazard mitigation, comprehensive land use and development planning [14].

The methodologies described above measure the risk to natural hazards in a city and the critical facilities are taken as an integral part of a measured indicator.

As a result, the vulnerability and risk analysis of a particular health facility is incomplete. Specific methods for assessing the risk of health facilities are few; among these, there is the Vulnerability Assessment of Health Care Facilities during Disaster Events (VAHCF) that measures the operational vulnerability of a health care facility during disaster events [15]. The guidelines of the Safe Hospital Program (SHP) developed by the Mexican Government (SEGOB) and the Mexican National System for Civil Protection (SINAPROC) define a "safe hospital" as a unity whose services remain accessible and functional at maximum capacity and within the same infrastructure immediately following a disaster [16]. The VAHCF and SEGOB methods consider the administration and organization conditions in health facilities to obtain a risk indicator. However, none of these two methodologies provides a level of vulnerability.

Vulnerability indexes in a particular model must consider the relative importance of each element involved as well as the interaction among these elements. Thus, a model system is a set of interrelated elements with a cohesion and purpose. The behavior of a system is understood as the dependence to the answers of the stimulus received [17]. An example is the fact that the response of a non-structural element to large earthquakes depends on the safety of the structural element and the safety of a structural element depends on proper maintenance and use of seismic technology in its components. These concepts when are applied to the measurement of vulnerability to seismic hazards make a convenient approach because it can extend the framework of the evaluation by including the different elements involved in the operation and safety of the facility and in the emergency response.

The methodology developed here represents a holistic approach to determine the seismic vulnerability in a hospital and uses numerical values weighted according to the importance of the elements considered. The estimated seismic vulnerability is the result of the combination of the structural, non-structural, functional, and social-administrative characteristics of the analyzed hospital.

2 Method

Our model was developed based on a system of indicators estimated using a theoretical framework (TF). This TF considers the occurrence of an earthquake of magnitude 8 or greater impacting Mexico City and is based on existing international standards and recommendations for risk management in hospitals [4, 6, 14, 16]. The model takes into account the following elements:

1. *Structural*: identifies those structural problems that will cause damage to the structural components of the installations (for instance, load-bearing walls, concrete slabs, architectural shapes, construction joints, overweight to the original construction).
2. *Non-structural*: those components that are not part of the resistant system of the structure but are crucial to the effective operation of the facility (for instance, medical equipment, doors, windows, walls, electricity, air conditioning, gas, emergency system or equipment, furniture).

3. *Functional*: includes the geographical location of the hospital's buildings as well as the use of the architectural spaces, conditions of local hygiene and security and the analyses of outdoor areas.
4. *Social-administrative*: analyzes the administration and organization in the health facility to mitigate the impact of a disaster and the existing emergency and evacuation plans.

As part of the methodology, the following definitions are considered:

- *Parameter*: components of any of the four *elements* (such as walls, roofs, furniture, equipment).
- *Failure*: Problem detected for each parameter (such as broken windows, unlocked baby cribs, lack or inadequate response plans, lack of regular building maintenance).

Ten parameters for each element (represented by letters from A to J) and 65 failures for each parameter with a previously assigned numerical value were considered [18].

2.1 Step 1: Design of the theoretical framework

2.1.1 Step 1.1. Determination of theoretical indicators

Theoretical indicators (TIs) are obtained according to the following equations (considering the worst vulnerability scenario in case of the occurrence of a large earthquake). The equations to estimate the structural theoretical indicator elements for any building system combined with steel (TIA_1) and for any other system (TIA_2) are the following:

$$TIA_1 = [VA_1 \times (EDTC + SH)] = 11.35 \tag{1}$$

$$TIA_2 = [VA_2 \times (EDTC + SH)] = 10.38 \tag{2}$$

where

$$VA_1 = \sum_{I=A}^{J} A_I = 12.62 \tag{3}$$

$$VA_2 = \sum_{I=A}^{J} A_I = 11.53 \tag{4}$$

$EDTC = 0.30$ and it represents the maximum numerical damage value according to the topology of construction. $SH = 0.60$ and it is the maximum numerical value assigned to the seismic hazard location (see section 2.1.3). On the other hand, VA is the theoretical vulnerability value of the structural element obtained from the sum of the numerical values given to each failure for each parameter represented by the second letter of each element [18]. Thus, the parameters of the structural element are represented by (AA, AB, AC,…, AJ) and the failures of AA by (AA_1, AA_2, AA_3,.. AA_N). The total number of failures is 65 independently of the fact that the number of failures in each parameter can be different. In this case, the sum of the assigned theoretical numerical value to each failure is 12.62 and 11.53 for VA_1 and VA_2, respectively.

The non-structural (*TIB*), functional (*TIC*), and social-administrative (*TID*) theoretical indicators are estimated by:

$$TIB = \sum_{I=A}^{J} B_I \times (EDTC + SH) = 19.54 \times 0.90 = 17.59 \tag{5}$$

$$TIC = \sum_{I=A}^{J} C_I \times (EDTC + SH) = 9.93 \times 0.90 = 8.94 \tag{6}$$

$$TID = \sum_{I=A}^{J} D_I \times (EDTC + SH) = 11.45 \times 0.90 = 10.31 \tag{7}$$

where B_I (19.54), C_I (9.93) and D_I (11.45) were obtained as described above for *VA* [18].

2.1.2 Step 1.2: Classification of damage

The following classification of damage was used [18]: DL= damage to life; DF= damage that limits the operability of the hospital; DO = damage that inhibits the optimal operation of the installation. The greatest damage is DL to which a weight equal to 0.50 was assigned. A weight of 0.40 was considered for DF because it generates less damage with respect to DL. A weight of 0.10 was assigned to DO because it is less important than the previous two kinds of damages.

2.1.3 Step 1.3: Weighing

Five weighing factors are considered [18]:

1. Classification of Areas (*CA*). Area A: Zone of critical importance to protect life. One hundred percent of these areas are required to operate during the emergency in the hospital. Thus, the numerical vulnerability measured for these areas is arbitrarily multiplied by 0.50. Area B: these areas are less important for the operation of the hospital. However, it was considered that at least 80% of these facilities must operate during an emergency by multiplying the obtained numerical vulnerability by 0.40. Area C: these areas are less important than Areas B and it is considered that at least 20% of these facilities must be functional in disaster situations. Thus, the measured numerical vulnerability for these areas is multiplied by 0.10.
2. Expected damage degree according to topology of construction (*EDTC*). The weight of this factor is determined by classifying the characteristics of the building material of the hospital. This classification varies from the most to the less vulnerable installations, that is, those facilities which have incorporated anti-seismic technology. Six categories are considered classified from A to F with A as the most vulnerable installation (with a numerical value of 5) and E and F as the less vulnerable [18,19] (with a numerical value of 1) (table 1).

Table 1: Weight of expected level of damage by topology of construction (EDTC).

Classification by topology of construction	EDTC	Weight
A	5	0.30
B	4	0.25
C	3	0.20
D	2	0.15
E Y F	1	0.10

3. Seismic hazard (SH). The Mexico City's building code [20] classifies the soil of the city in three seismic zones: Zone I (hill zone) is mainly basalt lava flows with little water content; Zone II (transition zone) is composed of sands with fairly high water content from eroding volcanic cones that surround the Valley of Mexico; Zone III (lakebed zone) consists of silt and volcanic clay sediments of the bed of the historic Lake Texcoco with high water content and exhibits a huge amplification of ground motion resulting in severe damage to the installations located in this area. For this reason, a weight of 0.60 was granted to Zone III. Because in Zone II the expected damage is less than in Zone III, a value of 0.30 was considered and to Zone I, 0.10 was assigned because this is the area with less probability of damage.
4. Weight assigned to elements of measured vulnerabilities. *VIA*, *VIB*, *VIC* and *VID* are the structural, non-structural, functional and social-administrative measured vulnerabilities, respectively. As part of our methodology, a weight calibration constant (*w*) was considered assigned to each element according to its importance with respect to others elements. In this work, *VID* is considered as the most important element for reducing vulnerability in a hospital. For this reason, a weight of 0.40 (*wd*) was assigned to this type of vulnerability. This is justified by the fact that actions for mitigating risk and reducing vulnerability depend on the decisions of the authorities and personnel of the hospital [4]. *VIA* is weighted with 0.30 (*wa*). This weighting scheme is based on the assumption that the security of the building depends on its structural stability [16]. The non-structural parameter was weighted with 0.20 (*wb*). This parameter is considered the third most important in the present model because damage in the non-structural parameter supposes substantial economic losses in any hospital [21]. The functional element was weighted with 0.10 (*wc*) because this element has the less influence in the installation's security [16].
5. The weights assigned to failures' numerical values are obtained by multiplying the weight of the type of expected damage (DL, DF, or DO) by the weight of the area where the failure is identified (A, B, or C) (table 2).

Table 2: Weight applied to damage failures in the elements.

Classification of damage	Weight applied to damage failures in areas A ($w = 0.50$)	Weight applied to damage failures in areas B ($w= 0.40$)	Weight applied to damage failures in areas C ($w= 0.10$)
Damage type DL ($w= 0.50$)	0.25	0.20	0.05
Damage type DF ($w= 0.40$)	0.20	0.16	0.04
Damage type DO ($w= 0.10$)	0.05	0.04	0.01

2.2 Step 2: Measurement of the vulnerability index (*VI*)

The numerical values of the structural (*VIA*), non-structural (*VIB*), functional (*VIC*), and social-administrative (*VID*) measured vulnerabilities are determined as follows:

$$VIA = \frac{VA \times (EDTC + SH + wa)}{TIA} \tag{8}$$

$$VIB = \frac{VB \times (EDTC + SH + wb)}{TIB} \tag{9}$$

$$VIC = \frac{VC \times (EDTC + SH + wc)}{TIC} \tag{10}$$

$$VID = \frac{VD \times (EDTC + SH + wd)}{TID} \tag{11}$$

2.3 Step 3: Estimation of the level of vulnerability (*LV*)

As a final step, with the measured values of *VIA, VIB, VIC*, and *VID*, the level of vulnerability is determined according to table 3.

A schematic summary of the procedures to follow after the TF has been established is shown in fig. 1.

To assess vulnerability following the methodological procedures described above, different tools are used [18]. Tool 1: Format to classify the analyzed hospital's areas; this allows the classification of the different types of areas in the health facility. Tool 2: Format to perform the structural assessment; with this tool the current condition of the structural system is identified as well as the age of construction, type of building system, architectural configuration problems, and past damages in the structural and non-structural architectural elements. Tool 3: Format to assess the interior areas; this allows the identification of the probable damages in each area of the hospital by classifying the failures according to the

Table 3: Descriptive of levels of vulnerability.

Descriptive	Level of vulnerability	Observations and recommended mitigation measures
Unacceptable	0.81 – 1.0	Re-structuring of buildings to warrant the safety and operability of the hospital is completely necessary. Most elements of the hospital are in very-high risk and structural collapse is possible.
Very high	0.61 - 0.8	The implementation of mitigation measures to diminish the level of vulnerability is necessary. The health facility is in very-high risk of losing operability and safety.
High	0.41 - 0 .6	The implementation of actions to reduce the vulnerability to an acceptable level is urgent to warrant safety and performance of the institution. The hospital is in high risk to suffer severe damage.
Moderate	0.21 - 0 .4	The hospital can suffer some damage affecting the operability of the installation.
Low	0.00 - 0 .2	Regular building maintenance and updating the response and mitigation plans are recommended.

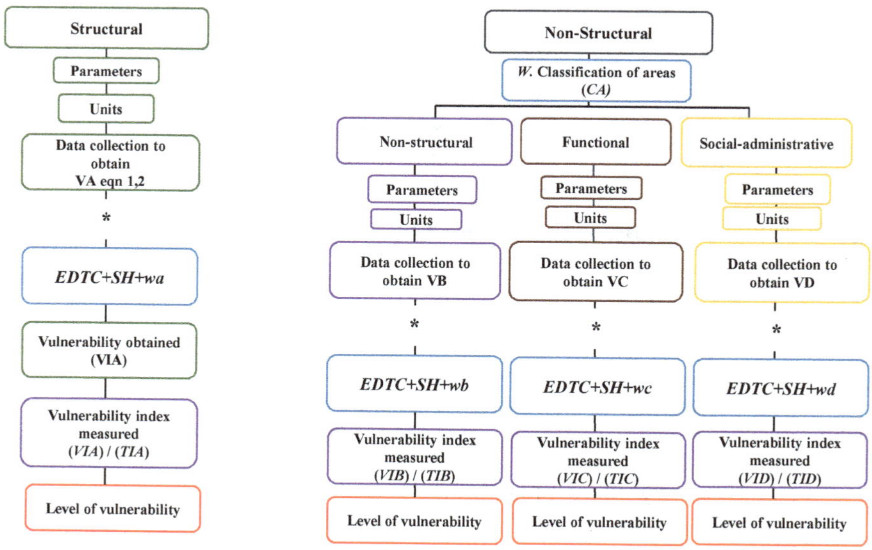

Figure 1: Schematic procedure to estimate the level of vulnerability.

TIA, *TIB*, *TIC* and *TID* parameters, eqns (1)–(2) and (5)–(7). Tool 4: Format to assess the outdoor areas; this tool detects those outdoor problems that increase vulnerability. Tool 5: Table describing and classifying the identified failures; this table classifies the problems identified that increase or decrease vulnerability and generates a complete report of potential and existing damages. Tool 6: A graph displaying the problems and possible damage identified; it displays the causes of the level of vulnerability and possible solutions to reduce it; it also identifies the failures that must be attended immediately and prioritize mitigation actions.

3 Results

The three hospitals selected for this study are located on the lakebed zone of Mexico City [12]. This area has high water content and exhibits a huge amplification of ground motion [1, 13]. Using the methodology described above, the results obtained in the Mexico Children's Hospital (ChH) are shown in detail. In the case of the JeH and the CaH, only a summary of the main results is presented.

3.1 The Children's Hospital (ChH)

The very high and high levels of structural vulnerability of ChH (fig. 2(a)), are mainly due to the buildings of this health facility being more than 70 years old and the characteristics of their construction do not incorporate earthquake resistant technology (table 4). Besides, the "Mundet" building, Federico Gomez building as well as the laundry and machinery installations have a high probability to be seriously damaged in case of being impacted by a large earthquake due to their architectural H and large rectangles shapes, the lack of regular maintenance as well as the geological setting where these buildings are located. Overall, the structural problems identified in this hospital represent a very high level of damage to life. The main reasons for the very high and high levels of structural risk are the lack of appropriate internal regulations for disaster prevention and the lack of sufficient economical resources for preventive actions such as the installation of earthquake-resistant technology. Among the socio-organizational factors that increase structural risk, the lack of risk prevention actions was identified. The proposed solutions for reducing the vulnerability at ChH are: 1) Reconstruct the laundry and machinery buildings and remove heavy objects from their roofs such as machines and heavy water containers; 2) Remove all pieces of the pipeline installation and the machinery that is not currently used; 4) Provide maintenance to finishes to prevent falling; 5) Waterproof walls and ceilings: incorporate earthquake resistant technology to guarantee the stability of the buildings.

The high risk due to non-structural vulnerability is associated to functional aspects (fig. 2(b)). The main problems found are: 1) Fire and explosion are possible due to spill of dangerous substances; 2) Heavy panels in ceilings and large windows with glass surfaces without protection to prevent breaking are inappropriate; 3) The emergency plan is not adequate for the three medical work shifts; 4) Incomplete installations and equipment necessary to provide medical care to patients in areas A and B; 5) Lack of emergency installation and equipment (such as fire installation or seismic alert).

Some proposed solutions to reduce ChH's non-structural risk are: 1) Obsolete equipment and material must be removed; 2) It is necessary to exchange needed heavy materials placed on the roofs of the installations and finishes by lightweight materials as well as to replace old equipment and installations (e.g., energy generators, elevators machines, washing machines) by new ones; 3) It is necessary to place emergency equipment, flexible joints in piping installations and use brackets on shelves and furniture to avoid downfall objects during a large earthquake.

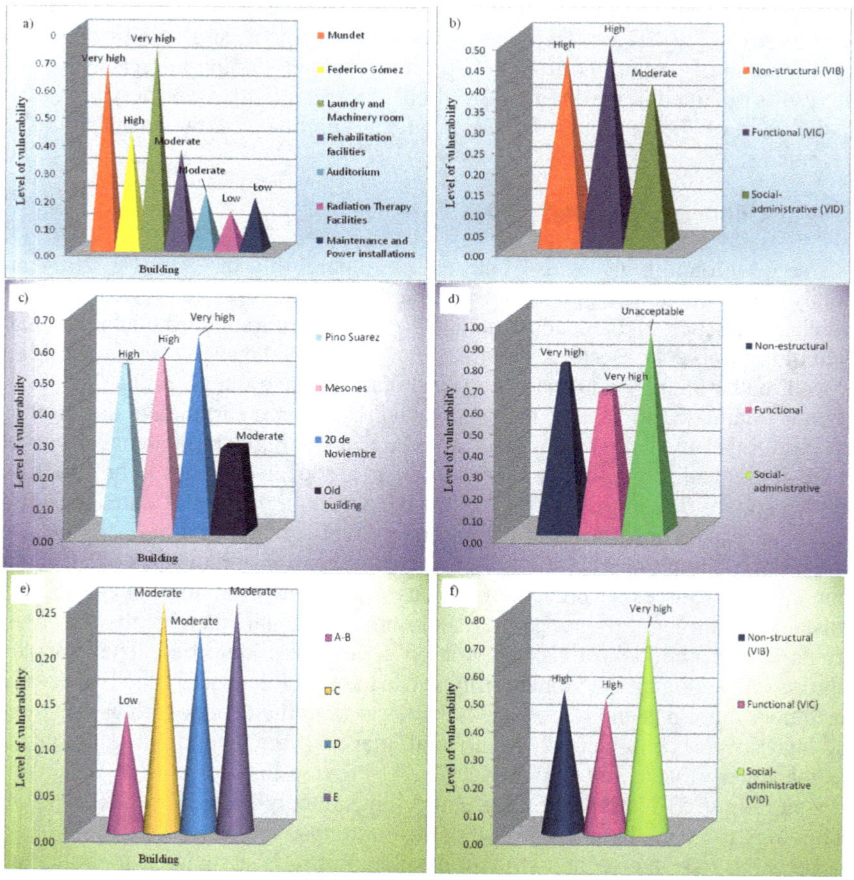

Figure 2: (a)–(b) Structural, non-structural, functional, and social-administrative vulnerabilities for the Children's Hospital; (c)–(d) the same for the Jesus Hospital; (e)–(f) the same for the Cardiology Hospital.

Table 4: Structural vulnerability in the Children's Hospital.

Building	(C2) VA[1]	(C3) Description TC[2]	(C4) Classification TC[3]	(C5) EDTC[3]	(C6) SH (Zone III)[4]	(C7) Calibration constant *wa*[5]	(C7) (C5+C6+C7)[6]	(C8) *IVA* (C2 * C7)	(C9) *ITA*[7]	(C10) *(C8/C9)*[6]	Level of vulnerability *LV*[8]
Mundet	6.21	Mixed. Reinforced concrete with brick walls	C	0.20	0.60	0.30	1.10	6.83	10.38	0.66	VERY HIGH
Federico Gómez	4.28	Reinforced concrete (RC) with minimum seismic design	D	0.15	0.60	0.30	1.05	4.49	10.38	0.43	HIGH
Laundry and Machinery Rooms	6.84	Mixed. Reinforced concrete with brick walls	C	0.20	0.60	0.30	1.10	7.52	10.38	0.73	VERY HIGH
Radiation therapy	3.41	Mixed. Reinforced concrete with brick walls	C	0.20	0.60	0.30	1.10	3.75	10.38	0.36	MODERATE
Auditorium	2.03	Reinforced concrete (RC) with minimum seismic design	D	0.15	0.60	0.30	1.05	2.13	10.38	0.21	MODERATE
Radiotherapy facilities	1.28	Mixed. Reinforced concrete with brick walls	C	0.20	0.60	0.30	1.10	1.41	10.38	0.14	LOW
Maintenance and Power installations	1.80	Mixed. Reinforced concrete with brick walls	C	0.20	0.60	0.30	1.10	1.98	10.38	0.19	LOW

[1] Equation (2); [2] See step 1.3 of Section 2; [3] See table 2; [4] See step 1.3 of Section 3; [5] See step 1.3 of Section 4; [6] See equation (8); [7] See equation (2); [8] See table 3.

The functional element has high level of vulnerability and risk (fig. 2(b)). The main problems identified are: 1) Lack of special rooms for patients in shock; 2) Lack of emergency exits and stairs; 3) The security and triage areas are not delimited. 4) Heavy equipment and water tanks placed on rooftops of some of its installations. The main solutions recommended are: 1) To provide areas A and B with appropriate equipment and installations; 2) Install emergency exits and stairs; 3) Establish security and triage areas.

The social-administrative parameter in ChH has moderate level of vulnerability (fig. 2(b)). The main problems found are: 1) The emergency response plan was developed only for the morning and afternoon's medical staff, do not include the night staff; 2) The list of staff and service providers is incomplete; 3) Blueprints of buildings are outdated; 4) Medical Staff is untrained on the use of triage cards to classify injured people during an emergency situation. As in the functional case, the main cause of these problems is the lack of awareness of them among the hospital's authorities. Probable solutions are: 1) Provide equipment (air conditioning, environment purifiers, oxygen and suction of liquids) and furniture for optimal medical care in areas A and B; 2) Prepare a list of available medical personnel and of supplies companies in case of emergencies and disasters; 3) Update plans and provide staff training for patient care in case of disaster situations.

3.2 The Hospital de Jesus (JeH)

For the Hospital de Jesus (JeH), the results show that the highest vulnerabilities are in the social-administrative, the non-structural and functional elements (fig. 2(d)). The socio-administrative element has unacceptable level because this hospital has not implemented its emergency plan in case of disasters yet. The very high levels of vulnerability of the non-structural and functional elements are mainly because of the lack of safety devices to protect equipment and medical furniture (fig. 2(d)). In general, this installation could suffer serious damage and loss of operability due to the impact of a large earthquake [22]. The structural element has very high level of vulnerability in the 20 de Noviembre building and two more were classified with high vulnerability level (fig. 2(c)). This is mainly due to the structural characteristics of the buildings and their architectural shape. The old building has not serious problems. However, it could suffer considerable damage in case of a major earthquake by the pounding of adjacent buildings. In this case, the losses would be very high because this installation has great historical value being the first health facility established in Latin America in the colonial period. Thus, it is considered important to incorporate anti-seismic technology and specific studies of structural engineering to reinforce its structure.

3.3 The Cardiology Hospital (CaH)

The CaH's seismic vulnerability is similar to that of the other two analyzed hospitals. This hospital suffered serious damage during the 1985 Michoacan earthquake [15]. As a consequence, all buildings that did not collapse, were reinforced with earthquake resistant-technology. For these reasons, the structural vulnerability of this hospital was identified between moderate and low (fig. 2(e)) [18, 22]. However, in this installation areas are located that provide life support to patients: surgery rooms, intensive care and laboratories. Because of these reasons, it is necessary to implement actions to reduce vulnerability to a low level to warrant the safety within these buildings. The non-structural and the functional parameters have high level of vulnerability (fig. 2(f)), indicating the need to reduce it to an acceptable level by implementing mitigation measures to protect the life of patients and personnel as well as the operability of the installation in case of emergencies. The social-administrative parameter was identified with very high level of vulnerability (fig. 2(f)). In this case, it is necessary to implement mitigation measures as well as a proper response plan to reduce this type of vulnerability to acceptable levels [22].

4 Discussion

Our methodology demonstrates the importance of detailed measurements to determine vulnerability in a health facility to prioritize actions to reduce the impact of a large earthquake. The observed very-high and high levels of structural vulnerability in our cases of study are mainly due to the lack of maintenance and technology to withstand seismic shaking as well as the use of complex architectural shapes. Other factors that increase the structural vulnerability are the lack of adequate separation among the hospital's installations generating the effect of pounding among these buildings in the case of the occurrence of a local earthquake.

Also, the type and age of construction as well as heavy objects or architectural elements placed on the roofs of the buildings not considered in the original structural design, increase the susceptibility of these buildings to suffering large damage by passing seismic waves.

Related to the non-structural elements, the main problem is the use of building materials not appropriate for hospitals such as heavy panels in ceilings and large windows with glass surfaces without protection to prevent breaking. The main problems identified in the functional parameter are the lack of brackets on shelves and furniture to avoid downfall objects as well as the need to install locks to beds, baby cribs and other furniture that could fall during a large earthquake. Others factors that increase the functional vulnerability are the lack of flexible material pipes for fluids and building's maintenance. The unacceptable level of vulnerability in the social-administrative element is caused by the lack of an emergency plan and of mitigation measures to withstand a large earthquake. It is suggested that it would be possible to reduce the risk and vulnerability in hospitals with the application of low cost mitigation measures such as development and implementation of a complete operational emergency plan, providing regular structural maintenance to buildings, removing heavy installation from the roofs of the structures, installing locks in the medical devices and equipment as well as implementing structural studies to incorporate anti-seismic architecture.

5 Conclusions

The methodology developed here allows the implementation and/or strengthening of existing programs to reduce the levels of structural, non-structural, functional, and social-administrative vulnerabilities of health facilities in Mexico City. It also identifies those actions needed for proper operation of these installations in disaster situations. The vulnerability problems found in the three cases of study are common among them even though they are having different structural, organization and medical specialties. For these reasons, it is suggested that it would be important for the Mexican government to develop a special program to determine the level of vulnerability to seismic hazards of at least the main and secondary health facilities in Mexico City by applying this method. Hospitals are only truly safe from disasters when they are functioning at maximum capacity immediately after a hazard strikes.

The JeH is the most vulnerable of the three cases of study with an unacceptable level of vulnerability in the social-administrative elements. The ChH has A very high level of vulnerability in the structural element. In the operational and the non-structural elements, this hospital has a high level of vulnerability. In general, the CaH was identified with the lowest level of vulnerability in the structural element. However, the social-administrative elements were measured with a very high level of vulnerability. For these three cases of study, it is necessary to develop a mitigation plan based on the results of this research. This will allow the significant reduction of their vulnerability in emergency and disaster situations.

Acknowledgements

The authors are grateful to G. Moran-Rodriguez for their assistance during the field research. Special thanks are due to Y. Bonilla-Moran, A. Moran-Balderas and G. Rodriguez-Quiroz for their support during this study. We also appreciate the collaboration of the authorities, medical staff, patients and workers of the Cardiology Hospital, the Children's Hospital and the Hospital de Jesus for their collaboration in obtaining the local data.

References

[1] Gunn, S.W.A., Health effects of earthquakes. *Disaster Prevention and Management*, 4(5), pp. 6–10, 1995.

[2] Hospitals Safe from Disasters, Reduce risk, Protect Health Facilities, Save Lives; United Nations, 2008–2009, World Disaster Reduction Campaign, Online.www.unisdr.org/wdrc-2008-2009

[3] Sabido, C.R., Padua, G.A., Rodríguez, M.H., Toledo, B. & Yanez, V.A., Efectos del terremoto del 19 de septiembre de 1985 en el Hospital General de la Ciudad de México, algunas consideraciones. *Salud Pública Mexicana*, 28(5), pp. 521–526, 1986 (in Spanish).

[4] Nateghi-Alahi, F. & Izadkhah, Y.O., Earthquake disaster management planning in health care facilities. *Disaster Prevention and Management*, 13(2), pp. 130–135, 2004.

[5] Fawcett, W. & Oliveira, C.S., Casualty treatment after earthquake disasters: development of a regional simulation model. *Disasters*, 24(3), pp. 271–287, 2000.

[6] Brandeau, M.L., Mccoy, J.H., Hupert, N., Holty, J.E. & Bravata, D.M., Recommendations for modeling disaster responses in public health and medicine: A position paper of the Society for Medical Decision Making. *Medical Decision Making*, 4(29), pp. 438–460, 2009.

[7] UNAM Seismology Group, The September 1985 Michoacan earthquakes: Aftershock distribution and history of rupture. *Geophysical Research Letters*, 13(6), pp. 573–576, 1986.

[8] Degg, M.D., Earthquake hazard assessment after Mexico (1985). *Disasters*, 13(3), pp. 237–246, 1989.

[9] Zeballos, J.L., Health aspects of the Mexico earthquake 19th September 1985. *Disasters*, 10(2), pp. 141–149, 2007.

[10] Soberon, G.J. & Sepulveda, J., The health care reform in Mexico: before and after the 1985 earthquake. *American Journal of Public Health*, 76(6), pp. 673–680, 1986.

[11] Moran-Rodríguez, S. & Novelo-Casanova, D.A., Spatial analysis of the level of exposure to seismic hazards of health facilities in Mexico City, Mexico. *Geomatics, Natural Hazards and Risk*, 3(3), pp. 279–292, 2012.

[12] Cardona, O.D. & Hurtado, J.E., Modelación Numérica para la Estimación Holística de Riesgo Sísmico Urbano, Considerando Variables Técnicas, Sociales y Económicas. http://desenredando.org/public/articulos/2001/estriesgo/EstimacionRiesgoSismicoUrbano-1.0.0.pdf (in Spanish).

[13] Indicadores para la Medición del Riesgo: Fundamentos para un Enfoque Metodológico, Programa BID/IDEA de Indicadores para la Gestión de Riesgos Web Site, Universidad Nacional de Colombia, Manizales, http://idea.unalmzl.edu.co (in Spanish).

[14] Lisa, K.F., Russell, W.J. & David, N.S., Community vulnerability assessment tool methodology. *Natural Hazards*, 3(4), pp. 163–176, 2002.

[15] Arboleda, C., Abraham, D., Richard, J. & Lubitz, R., Vulnerability assessment of health care facilities during disaster events. Infrastructure System, 15(3), pp. 149–161, 2009.

[16] Secretaría de Gobernación y Sistema Nacional de Protección Civil, Guía Práctica del Programa Hospital Seguro, SEGOB: México Distrito Federal, pp. 1–24, 2008 (in Spanish).

[17] Lantada, N., Irizarry, J., Barbat, A.H., Goula, X., Roca, A., Susagna, T. & Pujades, L.G., Seismic hazard and risk scenarios for Barcelona, Spain, using the Risk-UE vulnerability index method. *Bulletin of Earthquake Engineering*, 8(2), pp. 201–229, 2009.

[18] Moran-Rodríguez, S., Modelo cuantitativo para la estimación de la vulnerabilidad de las instalaciones de salud ante un evento sísmico, Tesis

doctoral, Universidad Nacional Autónoma de México: México D.F., pp.100–142, 2012 (in Spanish).
[19] Grünthal, G. (ed) European Macroseismic Scale 1998 Vol 15, Conseil de Léurope, Cahiers du centre Européen de Géodynamique et de Séismologie: Luxembourg, 1998.
[20] Gaceta Oficial del Distrito Federal, Reglamento de Construcción para el Distrito Federal, Reporte Enero de 2004, http://www.crid.or.cr/digitalizacion/pdf/spa/doc7292/doc7292.htm (in Spanish)
[21] Bitrán, D., Impacto Económico de los Desastres Naturales en la Infraestructura de Salud, CEPAL: América Latina y el Caribe, pp. 30–74, 1996 (in Spanish).
[22] Moran-Rodríguez, S. & Novelo-Casanova, D.A., Seismic risk of three main health facilities in Mexico City, Mexico. *WIT Transactions on the Built Environment*, 150, pp. 197–209, 2015.

Advances in urban planning and territorial management in the city of Lorca after the earthquake of May 11, 2011

S. García-Ayllón Veintimilla, A. Tomás Espín
& J. L. Ródenas Quiñonero
Technical University of Cartagena (UPCT), Spain

Abstract

Although the Lorca earthquake of May 11, 2011 had a moderate magnitude, it finally caused several casualties, being the first earthquake with deaths in Spain in the past 40 years, and important structural damages, forcing about 10,000 people to be unable to return to their houses. The city of Lorca is located in an environment of certain seismic vulnerability due to the presence of the Alhama fault, so the possibility of another earthquake cannot be ruled out. Nevertheless, management of the process and the mechanisms for damage assessment in urban areas are a topic which to date has scarce tradition in Spanish engineering. Therefore, the actions to be undertaken in the urban regeneration must be based on integral tools, such as seismic microzonation, and must affect certain fields of knowledge, such as territorial management and urban planning. This paper presents new techniques for aftershock management and damage assessment, with the experience acquired from the 2011 earthquake; it also analyzes the proposals that have been recently developed in the city of Lorca from the territorial point of view.
Keywords: earthquake management, urban microzonation, aftershock evaluation techniques, seismic territorial management.

1 The Lorca earthquake of May 11, 2011

The Lorca earthquake of May 11, 2011 had a magnitude of 5.1 M_w, being preceded two hours earlier by a premonitory earthquake of 4.5 M_w. The earthquake was also felt in the neighbouring provinces of Almeria, Albacete, Granada, Alicante and even in Jaén, Málaga, Ciudad Real and parts of the city of Madrid (fig. 1). Multiple aftershocks occurred in the hours following the shock of 18:47, with that of the largest magnitude of 3.9 M_w occurring at 22:37, causing finally 9 deaths and more than 300 injured.

(a) (b)

Figure 1: Location of Lorca in (a) Spain and (b) Murcia.

The Murcia region where Lorca lies is one of the most active seismic zones of Spain. Nevertheless, the Lorca basic acceleration of 0.12*g* in the Spanish seismic code [1] is not among the most dangerous (fig. 2).

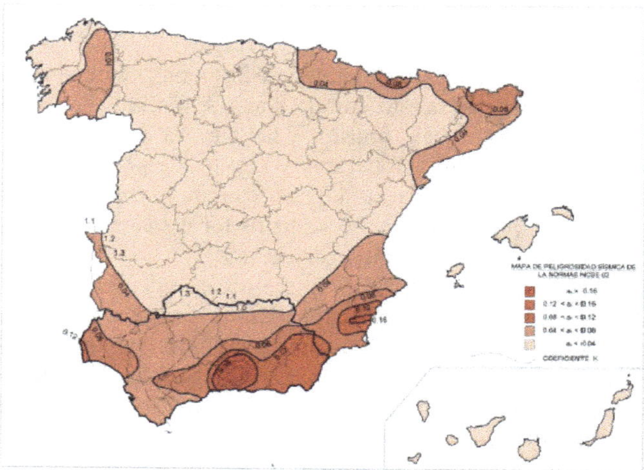

Figure 2: Map of seismic intensities in Spain [1].

The main quake occurred on land approximately 2 km northeast of the town of Lorca. Its hypocenter was very shallow, at a depth of about 1,000 metres. The area is located on the plate boundary between the Eurasian plate and the African plate. However, most of the earthquakes in the region do not often reach perceptible magnitudes and are not, therefore, usually felt. It has been estimated that the earthquake was a direct result of a strike-slip fault close to another major fault, the Alhama of Murcia one. This fault is a very shallow line, which extends to a length from 40 to 50 km, in whose vicinity local specialists noted the presence of rupture surfaces [2].

2 Damage assessment and zoning

The analysis of this earthquake was particularly interesting from the standpoint of damage. Many public buildings, housing and heritage monuments were affected. Within the zonal distribution of damage, certain areas not particularly close to the epicentre were notable, such as the neighbourhoods of La Viña and San Fernando or the old town.

The severity of the shock was largely due to the combination of shallow depth (an earthquake at a depth of one kilometre is very rare in the area) and a moderate magnitude. In Lorca, near the epicentre, the seismic event was assigned with an EMS-98 macroseismic intensity of I = VII, while other nearby areas detected movements of I = V [2].

An estimated 80% of homes were to some extent damaged, with over 1,300 having to be demolished in the following months. However, this distribution of damage was quite heterogeneous and upon analysis did not relate to only seismic parameters. In addition, it should be noted that despite the high number of homes affected, only one building actually collapsed during the earthquake (fig. 3).

With regard to aftershock damage assessment, from the very moment of the earthquake, a large and heterogeneous group of volunteer technicians were responsible for the emergency evaluation of the housing in Lorca, a task in which the authors participated. The object of this evaluation corresponds to an initial baseline assessment of safety and habitability of buildings, so the data obtained are approximate. A total of 7,839 buildings were analyzed during the first week. These data have been incorporated by the town hall and the Department of Public Works and Planning to the municipal mapping database to make available georeferenced damage data for analysis from a territorial perspective.

(a) (b)

Figure 3: (a) Church of Santiago; (b) building collapsed in La Viña.

The buildings were classified with green, yellow and red labels. The green classification meant safety of use, although the building could have minor damage such as cracks, crevices or small landslides. The yellow rating indicated a damaged building with access allowed based on the elimination of risk (e.g. the demolition of damaged roofs or parapets). Use of these buildings could also be cautiously granted as long as they had no more than non-decisive structural damage to their overall stability, following a visual inspection. The red label marked a structurally damaged building, with access to the property being prohibited. In fig. 4, typical damage identified by green, yellow and red labels are shown.

Globally, the damage in a total of 5,155 buildings was analyzed in the metropolitan area of Lorca. Of these, 889 (13% of the total) were initially classified to be at the yellow or red level. Damage is interesting in some census districts of La Viña neighbourhood, such as the 1013th and 1024th districts, with percentages of damaged buildings being the 41% and 40% of the total number of buildings of each district, respectively. Others, such as the 1004th census district, which is representative of the historic city centre, suffered damage classified as yellow or red in 16% of its buildings [2].

In this sense, a field study to document the seismic behavior in reinforced concrete buildings damaged in Lorca earthquake, 2011 has been implemented by the Department of Civil Engineering of Universidad Politécnica de Cartagena. After a comparison and validation process, from an initial sample of 1,050 buildings, a total of 406 homogeneous cases have been selected, removing unusual data or those where insufficient information was available. For every building, the observed damage grade μD in EMS-98 scale [3] has been reported.

Table 1 shows the age distribution of the analyzed buildings as a function of the construction periods, defined by considering the Spanish structural and seismic codes. Fig. 5 shows the observed damage grade depending on its period of construction. For example, within the group of structures constructed prior to 1963, 42.2% presented high damage levels ($\mu D = 3$), whereas this percentage diminished to 6.6% in the buildings constructed after 1996. On the contrary, only 6.7% of the buildings belonging to the first group presented a slight grade of damage ($\mu D = 1$), with this value reaching 39.6% in the more recently constructed buildings.

Nevertheless, as compared to the damage in buildings, the infrastructures were barely affected. The viaducts and tunnels of the A-7 road and the urban bridges of the city withstood the earthquake well and only minor damage was reported. In the railway infrastructures, the Lorca Sutullena station was the most affected and was demolished, while the Lorca-Águilas line (which itself was rendered unusable by flood damage months later) was barely damaged. Similarly, in water infrastructures such as the dams of Valdeinfierno and Puentes there were no problems, with the only damage being registered in the urban drainage networks.

In this way, from the first moment of the evaluation, it could be seen that the distribution of damage was very varied which meant that its explanation required territorial analysis of the urban environment. The asymmetric distribution of parameters such as the urban plot or the soil nature generated an unexpected distribution on many occasions, if a purely traditional seismological analysis was followed.

The town of Lorca, which currently has more than 91,000 inhabitants, sits on a narrowing of the River Guadalentín. It is an area where the river leaves the Cuenca

(a)

(b)

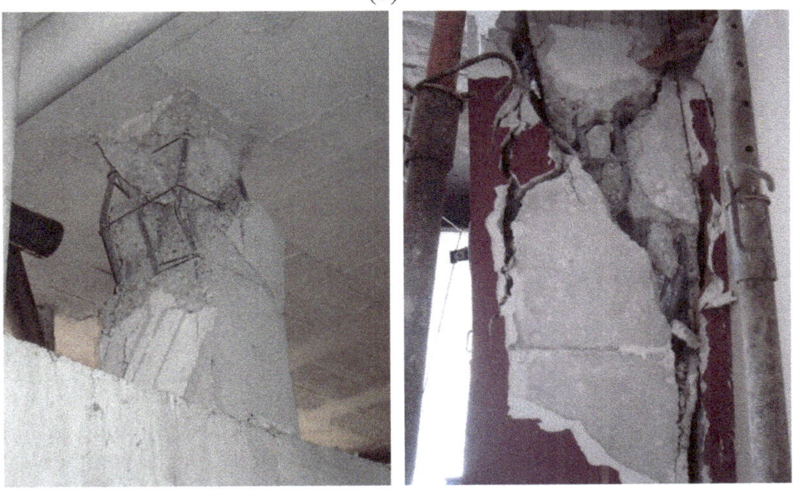

(c)

Figure 4: Damage in structural and non-structural elements of buildings after the Lorca earthquake, 2011 typically labelled as (a) green, (b) yellow or (c) red. (Source: authors.)

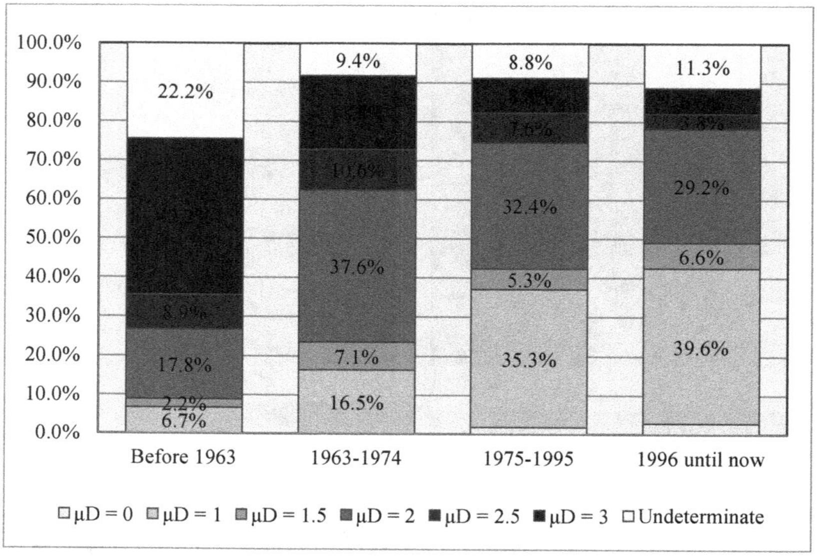

Figure 5: Distribution of EMS-98 mean damage grade observed on RC buildings after the Lorca earthquake, according to its period of construction.

Table 1: Distribution of building typologies analyzed in relation to the Lorca earthquake and its periods of construction according to the Spanish seismic code level.

Period of construction	Spanish seismic code	Code level	Number of buildings	% Buildings
I. Before 1963	-	Pre-code	45	11.1%
II. 1963–1974	Recommendation MV-101 (1962)	Pre-code	85	20.9%
III. 1975–1995	PGS-1 (1968) and PDS-1 (1974)	Low code	170	41.9%
IV. 1996 until now	NCSE-1994 and NCSE-2002	Medium code	106	26.1%

Alta through a pass in the Sierra de la Tercia and which has been used as a place to ford the river since ancient times; this has made it a privileged communication centre for centuries. The activity of the Alhama de Murcia fault has always allowed territorial conditions ideal for urban development, because of the high escarpment that provided the primitive inhabitants of the city with excellent visibility and security. This primitive urban planning has not been without problems. The 1674 earthquake caused extensive damage to the city but marked the beginning of an ambitious program of urban reconstruction, generating a very special baroque district that remains to this day, protected by a special protection plan [4].

This historic urban district consists mostly of buildings of masonry wall structures with internal elements of low seismic adaptive capacity. Its construction types are considered highly vulnerable due to their high mass and their low embedment between the wall planes.

Another of the most influential factors in the earthquake registered on a site is called the local effect, which consists in the amplification or attenuation of the ground motion as a function of the geological structure and topography of the urban environment. In this regard, design codes used to define an amplification parameter, depending on the characteristics of the ground, which makes it possible to distinguish four soil types according with a parameter *C* in NCSE-02, or five types of soil according with a parameter *S* in Eurocode-8 [5].

A simple interpolation of this parameter, based on the high number of existing geotechnical studies in the town of Lorca (fig. 6) allows the elaboration of a division into zones that explains certain results of the earthquake.

Figure 6: Inventory of soil amplification factors C in Lorca, according to NCSE-02 classification.

In this sense, the Department of Public Works and Planning of the Region of Murcia has commissioned a microzonation study of Lorca, which has been recently developed by the Technical University of Cartagena [6]. From these data, a map overlay of soil characteristics versus the vulnerability index or damage grade can be implemented (fig. 7).

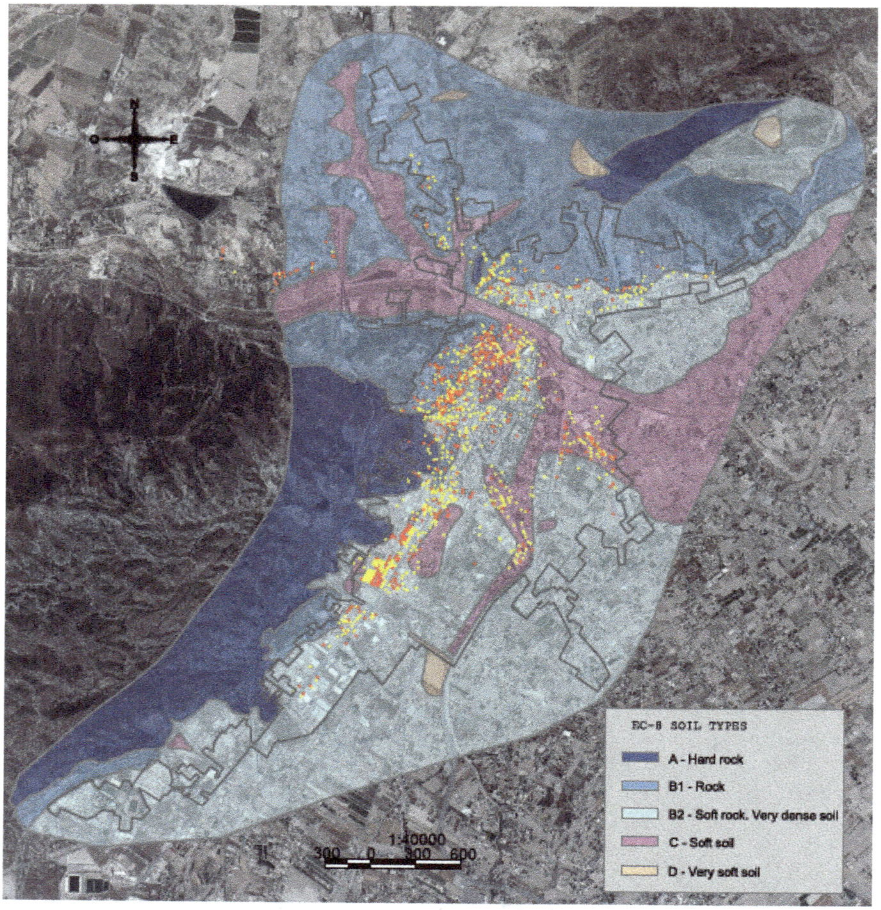

Figure 7: Overlapping of damage in buildings classified with yellow and red labels over soil types identified in microzonation study of Lorca, according to EC-8 classification.

An analysis of the incidence of damage out of the total number of buildings for each soil type shows a clear increase in damage to the soil type *C* (or type III in NCSE-02), referring to Guadalentín river sedimentary deposits. This explains cases such as that of the neighbourhood of San Fernando, where the 232 homes from the nine buildings in the neighbourhood had to be completely demolished because of the serious damage.

3 New evaluation techniques for information and aftershock management

An important aspect to consider after an earthquake is that damage evaluation provides homogeneous information of the affected area. There usually are, however, distorting variables (capacity of the evaluators, subjectivity, information management, etc.) as can be seen from acquired experience (table 2).

In the case of the Lorca earthquake no lives were lost due to the structural collapse of a building or infrastructure. Most of the material damage was associated with wrong urban configurations or little care in the construction details, with the most common structural pathologies being phenomena such as soft plant, short pillars or poor design in reinforced concrete.

Table 2: Damaged building classification categories.

	Turkey (1999)	Colombia (2009)	Chile (2010)
GREEN (safe, usable or habitable)	Little or no structural damage. Temporary occupation allowed.	Low risk for global stability, geotechnical problems, structural damage and / or non-structural damage.	No damage or very limited on the whole. No restrictions on use.
YELLOW (caution, restricted entry or temporary disablement, dubious safety)	Damage to structural and / or non-structural elements. Temporary occupation not allowed, except for emergencies.	Low risk, after measurements, global stability, geotechnical problems, structural damage and / or non-structural damage.	Damage, structural or non-structural sufficient for controlled risk to people. Partial closure and limited entry.
ORANGE (uninhabitable)	Not specified.	High risk for global stability, geotechnical problems, structural damage or non-structural damage.	Not specified.
RED (risk, unsafe or disablement)	Significant risk of collapse with aftershocks. Entry not allowed. Measures to prevent sudden collapse.	High risk for at least two (or very high for at least one) of the following reasons: global stability, geotechnical problems, structural damage, non-structural damage.	Elevated damage involving high risk of injury to persons. The building should be closed.

Based on the experience gained from this earthquake, a quick after-shock assessment form to ensure safety has been proposed by a workgroup formed by the Department of the Presidency of the Region of Murcia, the UPCT and other organisations and public institutions (fig. 8) [7].

ESTRUCTURAS PORTICADAS

EVALUACIÓN RAPIDA DE DAÑOS EXPEDIENTE Nº

IDENTIFICACIÓN DE LA EDIFICACIÓN

DAÑOS OBSERVADOS EN LA EDIFICACIÓN

VERDE

AMARILLO

ROJO

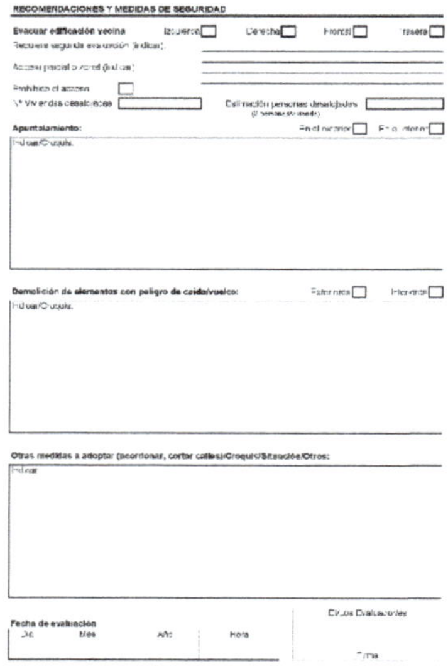
RECOMENDACIONES Y MEDIDAS DE SEGURIDAD

(a)

(b)

Figure 8: Completion of the proposed form: (a) front, (b) back [7].

For example, the building located in the Aprendices Square had an unexpected collapse during its demolition due to the damage on the ground floor and first floor (fig. 9). It had, according to its evaluator, only serious damage in facades and the fall of part of the breastplate housing, causing the financial ruin of the building. According to the proposed form, this building would be directly classified as a potential collapse (red label) since it presented foundation settlements and noticeably leaning.

(a) (b) (c)

Figure 9: Building in danger of collapse: evidence of failures in (a) joint between buildings and (b) corner pillar; (c) collapse.

4 The territorial management perspective in the analysis of the seismic vulnerability of urban areas

After the 2011 earthquake the Region of Murcia has launched certain planning measures in order to eliminate seismic vulnerabilities of its territory. In order to be prepared against a possible new earthquake, the SISMIMUR plan has been recently updated [8] by the General Directorate of Civil Protection of the Region of Murcia. This document specifies the seismic hazard, estimates vulnerability and the seismic risk in terms of damage and produces a catalogue of risk factors for buildings of particular importance that are located in areas where the intensity can be equal to or greater than VII for return periods of 475 years and 975 years in the Region of Murcia.

In the same direction, another interesting measure to qualify and improve the technical level of available staff for building design has been to start developing seismic strategy guides from the Ministry of Works and Planning of Murcia. These technical documents establish clear rules of seismic design for the reconstruction of buildings and provide rapid aftershock assessment protocols with standardized forms [9]. At a local level, a strategic action to prepare the city of Lorca for a possible new earthquake lies in adequate planning. Lorca is a city under reconstruction and in continuous urban growth (fig. 10).

In this sense, the above-mentioned microzoning of Lorca is very interesting, being implemented from a holistic and multidisciplinary perspective that is able to combine fields of knowledge such as seismic, geological, geotechnical or structural engineering. This is a standard procedure in cities with a seismic tradition such as Tokyo, Bogota or Santiago de Chile, but is a pioneering project for this size in Spain. It will also cover innovative areas such as territorial planning and land use, after detecting in the 2011 earthquake phenomena associated with topography, urban growth in undue environments risk or the urban configuration of buildings in an earthquake.

After the earthquake of May 11, there were some voices that postulated the possibility of modifying the existing seismic code (NCSE-02 for buildings) due to the acceleration of 0.369*g* achieved in an environment where the norm envisages just 0.12*g*, with the possibility of extending this to 0.196*g* under certain assumptions. The presence, however, of most damage associated with inadequate designs without any seismic quality and the lack of regard for any of the recommendations of the standard itself in effect denies the need.

The problem therefore lies rather in the need to adapt the local factor of each environment rather than to conduct an overall review and update of the state regulations after the occurrence of specific events. In many cases, the local effect is dominant on the effects of the source and wave propagation, and thus, its inclusion in the estimation of the expected ground motion is of prime importance.

Quantifying the local effect is very complex because it involves many aspects. In studies at regional scale such as SISMIMUR, more or less simple geological classifications tend to be employed, encompassing different soil types in the region of study. For each class an average amplification factor is adopted, which is

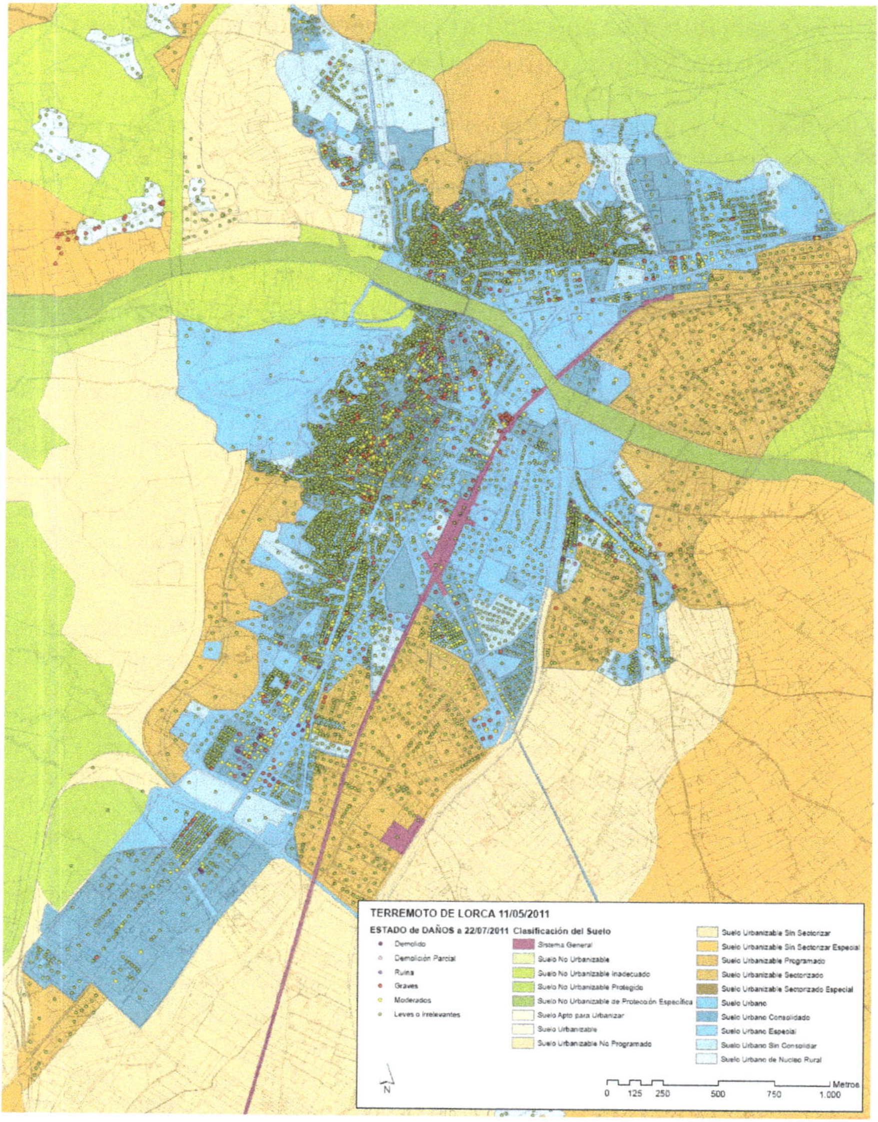

Figure 10: Distribution of damage in urban planning schedule.

applied to the estimated motion in hard rock or soil, thus the resulting movement includes the local effect [8].

In this field, it is important to consider the necessary adaptation of many buildings now to be rebuilt with CTE update legislation [10]. This will alter the original configuration of almost all these buildings. That presents a unique opportunity to review the entire General Plan, the principal technical tool in the Spanish legal system at municipal level.

A new more technological general plan with urban growth and development, taking into account the territorial results of performed microzoning would reduce as far as possible the seismic vulnerabilities of the city with the proven risk of Lorca and it would be a very useful and innovative tool.

5 Conclusions and future work

The experience provided by the Lorca earthquake of May 11, 2011 leads to the following conclusions:

1) The implementation of a planning and seismic strategy with fast action protocols is necessary. These protocols should incorporate damage assessment forms, which are rigorous yet simple and adaptable to a massive intervention by technicians and volunteers.

2) It is very important to integrate the local seismic factor. It is not necessary to impose general changes in seismic regulations based on specific events, but to update tools that empower, at regional and local level, the prevention of seismic vulnerabilities in the various municipalities.

3) In spite of the technical advances in seismic structural design, many urban areas still present a major level of seismic risk due to the high vulnerability of their buildings, mainly certain constructions designed without seismic regulations or based on out-of-date codes. Therefore, the period of construction is another of the main factors to be considered in order to plan the development of the urban plot and management of new emergency situations.

4) In this sense, the large number of buildings that will require modification of their configuration during the rebuilding because of the need to adapt to current building regulations recommends a complete overhaul of the General Municipal Plan in Lorca. Engineering urban planning, adapting building typologies to particular seismic problems based on tools such as microzoning or SISMIMUR in cities with high risk of major seismic events or vulnerabilities like Lorca, is recommended as a future policy.

References

[1] Ministry of Development of Spain, *National construction seismic code (NCSE-02)*: Madrid, Spain, 2002.

[2] Geographical National Institute of Spain, *Informe del sismo de Lorca del 11 de mayo de 2011*: Madrid, Spain, 2011.

[3] Grünthal, G. (ed.), *European Macroseismic Scale 98 (EMS-98)*, Conceil de L'Europe, Cahiers du Centre Européen de Géodynamique et de Séismologie: Luxembourg, 1998.

[4] Council of Lorca, *General Urban Plan of Lorca*: Lorca, Spain, 2007.

[5] Comité Européen de Normalisation, *Eurocode-8: Design of structures for earthquake resistance (EC-8)*: Brussels, Belgium, 2004.

[6] Department of Public Works and Planning of the Region of Murcia, *Estudio de microzonificación sísmica en el municipio de Lorca*, Universidad Politécnica de Cartagena: Cartagena, Spain, 2014.
[7] General Directorate of Civil Protection, *Evaluación rápida de daños en emergencias. Protocolos de activación y actuación del grupo de evaluación de daños*, eds. A. Frías & J.A. Guzmán, Department of the Presidency of the Region of Murcia: Murcia, Spain, 2015.
[8] General Directorate of Civil Protection, *Plan Especial de Protección Civil ante el riesgo sísmico en la Región of Murcia (SISMIMUR II)*. Department of the Presidency of the Region of Murcia: Murcia, Spain, 2015.
[9] Department of Public Works and Planning of the Region of Murcia, *Guidelines for the definition of an antiseismic strategy*, FHECOR Knowledge: Murcia, Spain, 2012.
[10] Ministry of Housing of Spain, *Technical Construction Design Code (CTE)*: Madrid, Spain, 2010.

Tsunami flood risk prediction using a neural network

H. Gotoh & M. Takezawa
Nihon University, Tokyo, Japan

Abstract

The earthquake off the Pacific coast of Tohoku (known as the "Sendai Earthquake") occurred on March 11, 2011 and severely damaged the Tohoku coastal area facing the Pacific Ocean. The number of casualties in the Sendai earthquake, as of August 9, 2013, included 15,883 deaths, with 2,656 people still missing. The cause of death for nearly 90% of those killed was drowning because many people were swept away by the resulting tsunami. Focusing on the area inundated by the current induced by the tsunami, even in adjacent areas, a difference was apparent in the inundation distances from the coastline. Clarification of the reason for this difference is necessary for city planning when considering disaster reduction in the future. In this study, the factors that promote inundation in the Sendai area are first clarified, excluding the effects from small ground undulations, using a neural network to establish a complex relationship between inundation and each of the factors. Based on the results on the factors that promote inundation, Iwata City in the Shizuoka Prefecture has been selected as a study area because its region is similar to the Sendai area that has flat level ground and also it is likely to be struck by a tsunami generated from any Tokai, Tounankai, and Nankai type earthquake. In the present risk prediction, the risk for tsunami damage in Iwata City is estimated and measures to reduce this damage are proposed.
Keywords: tsunami, neural network, "inundation factors", risk analysis.

1 Introduction

The Great East Japan Earthquake occurred on March 11, 2011, and the coastal zones along the Pacific Ocean in the Tohoku area were inundated by the current induced by the resulting tsunami. There was extensive damage and loss of life in an instant; however, a difference was seen in the inundation area away from the coastline. In cases of extreme inundation in the Sendai plain, inland areas 5 km from the coastline were flooded. Although it is assumed that objects along the coastline, including houses, roads, rivers, cultivated land and coastal forestation inhibited the tsunami-induced current, the effect of these objects on inhibiting this current has yet to be clarified.

In this study, based on the damage after the Great East Japan Earthquake, the factors that inhibit or promote inundation from tsunami-induced currents were investigated using a neural network. From the results, the potential damage caused by a tsunami from the Tokai earthquake, which is predicted to occur in the near future, was estimated.

2 Damage from tsunami due to the Great East Japan Earthquake

The Tohoku area along the coast facing the Pacific Ocean was severely damaged by the tsunami caused by the Great East Japan Earthquake. The wave height of the tsunami was reported to be between 2 and 21 m based on field investigations by numerous researchers [2]. Moreover, as shown in fig. 1, the value of the maximum run-up height was reported to reach nearly 40 m [2]. Approximately 90% of the casualties from the Great East Japan Earthquake were a result of the tsunami [3]. Fig. 2 shows examples of damage caused by the tsunami.

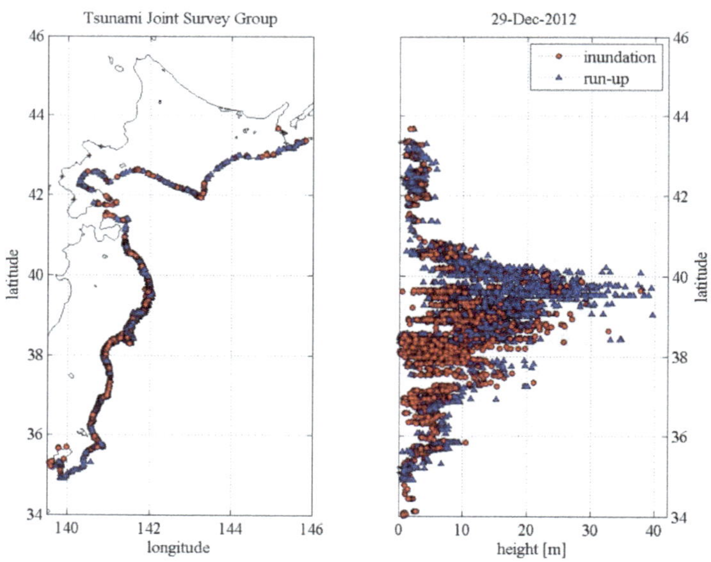

Figure 1: Tsunami wave height and run-up height measured by the 2011 Tohoku Earthquake Tsunami Joint Survey (TTJS) Group [3].

(a) Damaged building in Onagawa City

(b) Flotsam in Ishinomaki City

(c) Damaged hotel in Taro City

(d) Collapsed dike in Taro City

Figure 2: Examples of damage caused by the tsunami.

Fig. 3 shows the area in the Sendai plain inundated by the tsunami-induced current. The maximum distance of the inundation area from the coastline was reported to be approximately 5 km. However, as shown in Figure 3, there is a difference in the distances of the inundation area from the coastline, even if the area is primarily flat. It is assumed that the inundation area is affected by objects on the ground (predictor variables).

3 Methodology

3.1 Estimation of factors that promote inundation

To investigate the effect of predictor variables, a flat area of land was selected – the Sendai plain. As shown in fig. 4, 60 survey lines perpendicular to the coastline were set up at 200 m intervals. Along each survey line, the width of the beach, the width of coastal forestation, the width of fields including the ground (e.g., cultivated land), roads and parks, the width of big buildings such as factories and warehouses, the width of residential areas, the width of rivers perpendicular to the coastline, the width of rivers parallel to the coastline, the width of lakes and ponds and the wave height were selected as predictor variables to perform the analysis using a neural network. Fig. 5 shows examples of these objects. As shown in fig. 6, “the rivers perpendicular to the coastline” were defined as ranging from 0 to 45°

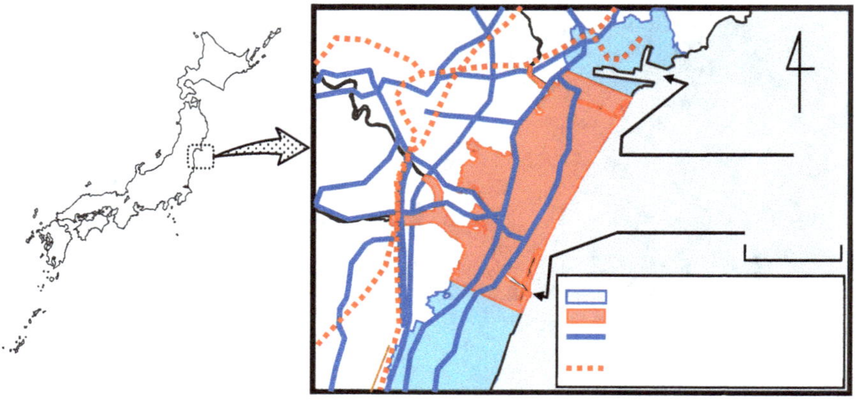

Figure 3: Map of the study area (Sendai plain).

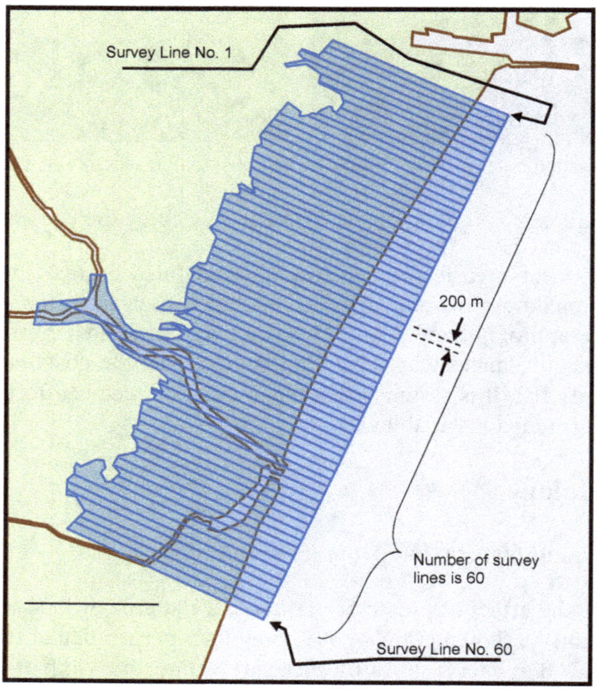

Figure 4: Survey lines in study area.

(a) Beach

(b) Coastal forestation

(c) Field (cultivated land)

(d) Big building

(e) Residential area

(f) River (canal)

Figure 5: Examples of predictor variables.

from the survey line and "the rivers parallel to the coastline" were defined as ranging from 46 to 90° from the survey line. The values of the predictor variables were measured as length across the survey line, and the values of the objective variable were equal to the inundation distance from the coastline along the survey line, as shown in fig. 7. The degrees of importance of the predictor variables (the factors that promote inundation) against the objective variable were estimated using a neural network. A neural network is a computational model based on the structure and functions of biological neural networks. Information that flows through the network affects the structure of the neural network because it changes, or learns, based on inputs and outputs. Neural networks are considered nonlinear statistical data modeling tools where the complex relationships between inputs and outputs are modeled or patterns are found. Here, an IBM software package, SPSS Neural Networks V 20, was used to create the hierarchical neural network.

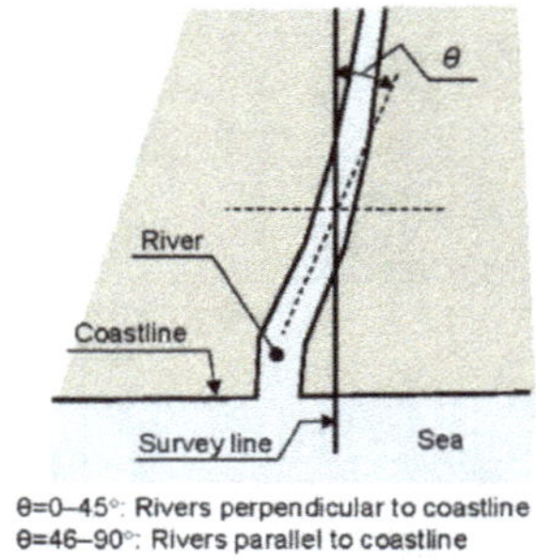

Figure 6: Definition of rivers perpendicular or parallel to the coastline.

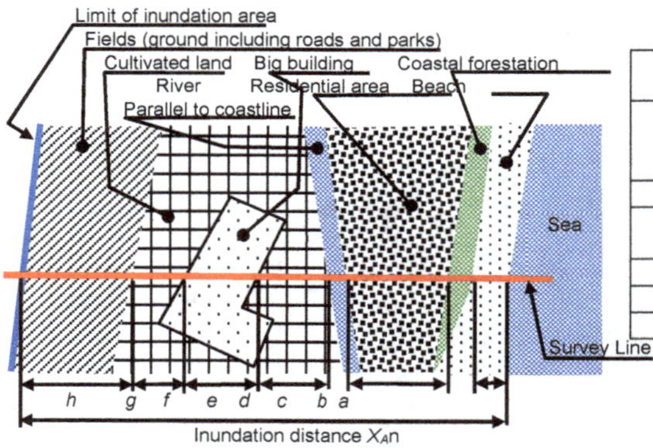

Predictor variables	Measured distance (m)
Fields (ground including roads and parks)	*h*
Cultivated land	*e+g*
River parallel to coastline	*d*
Big buildings	*f*
Coastal forestation	*b*
Beach	*a*

Figure 7: Example of method for measuring predictor variables.

3.2 Risk analysis for future tsunami

A risk analysis was conducted on the inundation due to a future tsunami caused by the Tokai, Tounankai, and Nankai type earthquake. The coast of Iwata City in Shizuoka Prefecture facing the Pacific Ocean was selected as the study area, as shown in fig. 8. A broad plain stretches out from coastline to inland, similar to the Sendai plain. It is estimated that a large earthquake will strike this area in the future, and the resulting tsunami will probably inundate towns along the coast. This area was divided into nine regions, each 3000 m long by 500 m wide, as shown in fig. 8. The component ratio of land use was surveyed using aerial photographs and a field survey. Then, the degree of risk for each region was assessed.

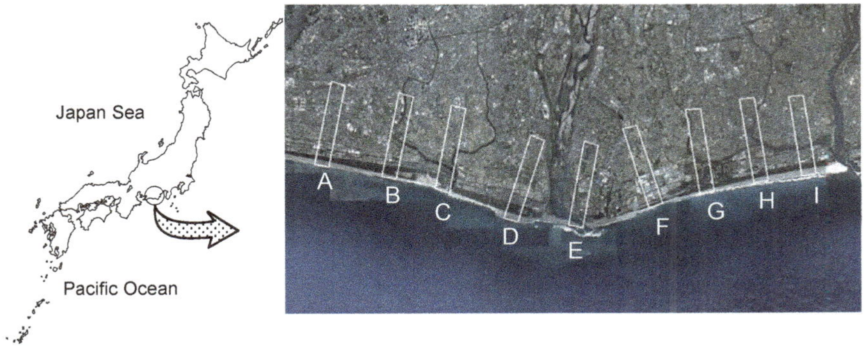

Figure 8: Test sites for tsunami risk analysis (Iwata City in Shizuoka Prefecture).

4 Results of analysis

4.1 Factors that promote inundation

The objective variable and predictor variables were measured using aerial photographs and survey results of the inundation area [4]. Table 1 shows the data (predictor variables) input into the neural network. Fig. 9 shows the calculated values using the neural network compared with the actual values (i.e. teaching data) X_A on the inundation distance from the coastline. As shown in fig. 9, the calculated values X_E are almost equal to the actual values, within ±3% error. The neural network estimated the inundation distance very well. Fig. 10 shows the degree of importance of the predictor variables against the objective variable (inundation distance) based on the calculations using the neural network. As shown in fig. 10, the inundation was promoted by the ground (including cultivated land), roads, parks and rivers perpendicular to the coastline. On the other hand, the inundation was inhibited by rivers parallel to the coastline, lakes and ponds, coastal forestation and beaches. In addition, although the inundation distance is considered to be strongly dependent on wave height, it had a low degree of importance in the present network because it was considered that there was little difference in wave height along the survey lines.

4.2 Risk analysis for a future tsunami

Table 2 shows the component ratio of land use in the nine selected regions. Table 3 shows the results of risk analysis based on fig. 10. As shown in table 3, the degree of tsunami risk differs among neighbouring regions. Measures to reduce the damage caused by a tsunami could include management of coastal forestation, creation of embankments and construction of rivers (or canals) parallel to the coastline for irrigation of fields for common use. It is also recommended that the foundations of arterial roads parallel to the coastline be made higher.

Table 1: Measured values for predictor variables.

Survey line No.	Beach (m)	Coastal forestation (m)	Field (m)	Cultivated land (m)	Residential area (m)	Big buildings (m)	Rivers parallel to coastline (m)	Rivers perpendicular to coast line (m)	Lakes and ponds (m)	Wave height (m)
1	65	305	2816	0	0	696	0	0	0	9.0
2	150	70	2144	0	110	1050	0	0	369	8.8
3	169	0	2693	0	300	426	0	0	387	8.6
4	262	0	2367	150	655	512	0	0	93	8.4
5	258	0	2006	25	1339	281	0	0	122	8.8
6	63	0	1008	0	1878	90	0	1465	0	8.7
7	185	225	1475	0	0	0	0	3117	0	8.6
8	185	220	169	2032	835	231	46	0	0	8.8
9	153	254	192	2029	884	105	38	0	0	12.6
10	166	197	672	1253	321	198	37	0	0	7.1
11	157	353	185	2152	0	0	43	0	0	10
12	147	436	182	1704	339	0	56	0	0	7.5
13	170	311	319	2245	155	0	46	0	0	7.3
14	174	285	300	2518	203	0	26	0	0	9.0
15	163	386	271	2376	769	45	29	0	0	10.2
16	174	664	106	2939	174	0	26	0	0	12.1
17	176	670	214	3035	0	0	24	0	0	13.8
18	163	574	0	3115	0	0	22	0	228	14.3
19	176	594	0	3204	120	0	22	0	0	11.5
20	156	540	0	3271	0	0	23	0	0	5.4
21	177	587	0	3132	0	0	23	0	145	5.4
22	165	709	159	2462	0	238	23	0	287	6.1
23	152	548	380	2589	76	120	21	0	0	6.1
24	160	559	155	2503	304	0	20	0	0	5.7
25	175	665	0	2790	0	0	27	0	0	5.3

Table 1: Continued.

Survey line No.	Beach (m)	Coastal forestation (m)	Field (m)	Cultivated land (m)	Residential area (m)	Big buildings (m)	Rivers parallel to coastline (m)	Rivers perpendicular to coast line (m)	Lakes and ponds (m)	Wave height (m)
26	241	297	368	2475	93	0	53	0	0	4.8
27	248	51	222	2685	503	111	27	33	0	5.3
28	206	63	66	100	1197	0	30	0	0	12.2
29	197	87	163	2934	674	0	30	0	0	11.2
30	200	488	154	2866	317	0	27	0	0	9.2
31	127	422	0	3029	337	0	33	0	0	7.2
32	114	614	0	2867	358	0	326	0	0	5.9
33	97	599	120	2485	374	0	30	0	0	5.8
34	100	603	0	2956	0	0	29	0	0	5.6
35	89	253	289	2058	0	0	0	951	0	5.3
36	82	641	110	2440	281	0	46	0	0	5.0
37	64	637	295	2283	425	0	44	0	0	4.9
38	58	661	136	2725	229	0	40	0	0	5.0
39	62	615	0	2997	0	0	44	0	0	5.2
40	49	622	127	2417	410	0	46	0	0	5.1
41	42	600	289	2432	184	0	46	0	0	5.3
42	42	613	109	2108	296	0	43	0	0	6.3
43	57	470	245	2292	141	0	91	0	0	16.1
44	57	344	176	2089	346	0	48	0	189	11.9
45	59	319	274	2777	72	0	30	0	185	8.0
46	55	331	191	1831	670	0	225	0	0	4.6
47	78	285	272	2618	486	0	238	0	0	3.8
48	75	307	222	3509	187	0	241	1338	0	4.1

Table 1: Continued.

Survey line No.	Beach (m)	Coastal forestation (m)	Field (m)	Cultiva-ted land (m)	Residen-tial area (m)	Big buildings (m)	Rivers parallel to coastline (m)	Rivers perpen-dicular to coast line (m)	Lakes and ponds (m)	Wave height (m)
49	101	439	398	2265	0	0	237	117	0	4.3
50	118	145	314	3112	0	0	303	256	0	4.5
51	106	50	442	1808	556	0	344	278	0	4.7
52	120	55	765	916	294	0	400	839	0	5.1
53	184	0	0	1348	0	0	402	1238	0	5.6
54	298	0	310	1383	0	74	0	1283	0	6.8
55	0	0	0	1956	1020	0	0	776	0	8.5
56	0	0	168	1986	1041	0	34	0	755	9.3
57	0	50	222	2171	754	0	43	0	180	8.4
58	287	0	273	2441	624	0	36	0	187	7.5
59	226	87	342	1977	715	91	33	0	218	10.2
60	203	121	0	2433	338	0	48	0	276	8.3

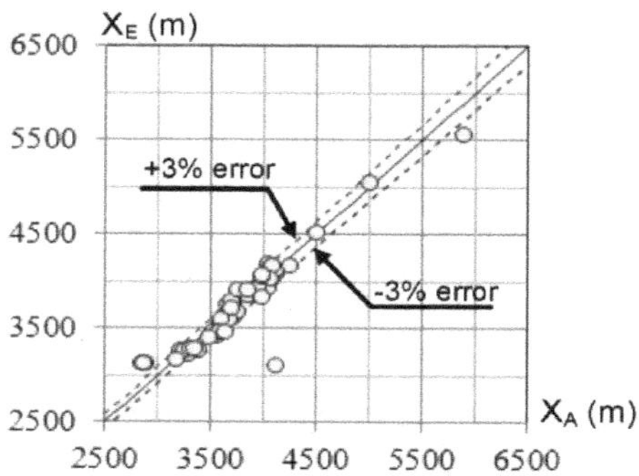

Figure 9: Comparison of estimated and actual values.

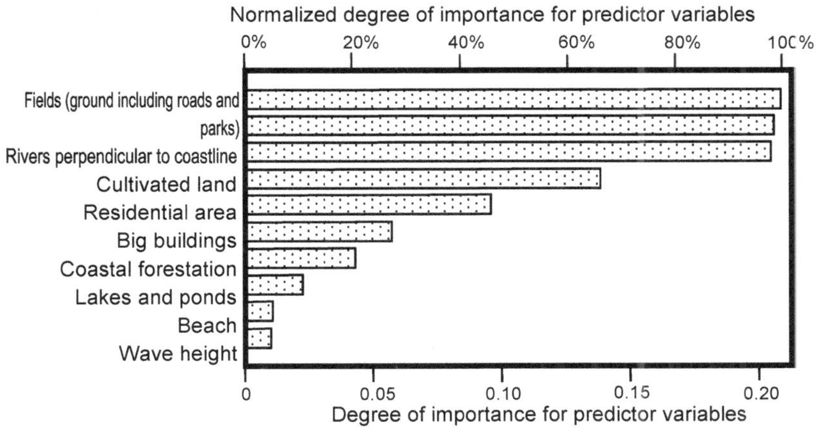

Figure 10: Factors affecting inundation area.

Table 2: Component ratios of land use in selected regions.

Predictor variables	Survey Area (%)								
	A	B	C	D	E	F	G	H	I
Beach	5.8	3.5	5.7	8.6	5.4	-	3.0	3.5	7.9
Coastal forestation	8.0	13.2	17.5	10.8	1.8	5.5	20.0	14.1	6.4
Fields (ground including roads and parks) /Cultivated land	54.7	53.7	52.1	56.6	68.5	75.9	57.4	52.3	33.2
Residential area	28.4	12.2	9.3	23.2	8.8	7.6	8.0	12.6	46.0
Big buildings	3.6	4.9	4.7	0.8	3.3	11.0	-	5.6	3.6
Rivers parallel to coastline	-	-	-	-	-	-	2.1	1.5	2.9
Rivers perpendicular to coastline	-	12.5	10.6	-	12.1	-	4.3	10.4	-
Lakes and ponds	-	-	-	-	-	-	5.2	-	-

5 Conclusion

In the present study, an attempt was made to clarify the factors that promoted (predictor variables) inundation in coastal regions from the tsunami after the Great East Japan Earthquake on March 11, 2011. Using risk analysis of these factors, the degree of risk from tsunami in Iwata City, Shizuoka Prefecture was also estimated and measures to reduce potential damage from a tsunami were proposed. The main results are summarized as follows:

- Using a neural network for the analysis, the predictor variables to reduce the inundation distance are rivers parallel to the coastline, lakes or ponds, coastal forestation, large buildings such as factories and warehouses and beaches. Moreover, the risk analysis for tsunami in the case of Iwata City in Shizuoka Prefecture was performed using the factors that inhibit inundation and the judgments on safety were made.

- For areas where earthquakes may occur in the future, managing coastal forestation, creating embankments and constructing rivers (or canals) parallel to the coastline are proposed as measures to reduce tsunami damage.

Table 3: Result of risk analysis for Iwata City in Shizuoka Prefecture.

Predictor variables	Survey area								
	A	B	C	D	E	F	G	H	I
Fields (ground including roads and parks)/ cultivated land	IV	IV	IV	IV	V	V	IV	IV	IV
Rivers perpendicular to		III	III		III		III	III	
Big buildings						II			
Coastal forestation		II	II	II			II	II	
Lakes and ponds							I		
Beach	II		II	II	II				II
Rivers parallel to coastline							I	I	I
Residential area	Exist (near coastline)	Exist (near coastline)		Exist (near coastline)				Exist (far from coastline)	Exist (near coastline)
Judgment on Safety	×	×	O	×	Δ	O	O	Δ	×

I: It can be expected to reduce damage greatly.
II: It can be expected to reduce damage.
III: Not exceeding 30% land use is not expected to reduce damage.
IV: 30–60% land use is not expected to reduce damage.
V: Exceeding 60% land use is not expected to reduce damage.
×: The area has large risk for loss of life due to tsunami-induced current.
Δ: The area has low risk for loss of life, although there is inundation by the tsunami.
O: The area has little risk for loss of life because of reduced inundation.

References

[1] Reconstruction Agency: Situation of damage for every prefecture, http://www.npa.go.jp/archive/keibi/biki/higaijokyo.pdf, (last accessed 2014.2.6).

[2] Coastal Engineering Committee of JSCE (The 2011 Tohoku Earthquake Tsunami Joint Survey Group): The 2011 off the Pacific coast of Tsunami Information, http://www.coastal.jp/ttjt/ (last accessed 2014.2.6).

[3] Cabinet Office, Government of Japan: White paper, http://www.bousai.go.jp/kaigirep/hakusho/h23/bousai2011/html/zu/zu004.htm (last accessed 2014.2.6).

[4] Geospatial Information Authority of Japan: General inundation area maps for the 2011 off the Pacific coast of Tohoku Earthquake Tsunami. http://www.gsi.go.jp/ (last accessed 2014.2.6).

Landslide risk management in Malaysia

C. H. Abdullah
Slope Engineering Branch, Public Works Department, Malaysia

Abstract

Malaysia is located in a tropical region where rainfall is abundant. Due to this condition, Malaysia experiences frequent flooding and landslides. The Slope Engineering Branch of the Public Works Department of Malaysia is given the task by the Malaysian Government to monitor and enhance slope safety. Losses due to landslides in Malaysia are estimated to be more than USD 1 billion since 1973. One of the first thing carried out by the Slope Engineering Branch after its establishment in 2004 is the drafting of a National Slope Master Plan (NSMP) to pave the way for enhancement of slope management. The goal of the NSMP is to reduce risks and losses due to landslides. There are ten components in the NSMP. The NSMP laid out strategies and action plans required to better manage the slopes in Malaysia. Some of the action plans that have been translated to practice include but are not limited to the following activities: inventorizing slopes; producing hazard and risk maps; carrying out public awareness and education programs; and forming committees among agencies to formulate effective slope mitigation policies and cooperation. The paper describes the actions taken by the Public Works Department to mitigate landslides in Malaysia.
Keywords: slope master plan, risks, losses, action plans, landslide, mitigation.

1 Introduction

Malaysia has been experiencing rapid economic development since the early 1990s. As a result, more hilly terrain areas have opened up for development. From 1973 to 2011, there have been a number of major landslides in Malaysia with a total loss of lives amounting to more than 600. Fig. 1 shows the number of landslide events and fatalities from 1973 to 2011. It is estimated that economic losses due to landslides since 1973 is more than USD 1 billion. Faced with landslides that seemed to increase annually, the Malaysian Government took steps to come up with mitigation measures. In 2004, the Slope Engineering Branch was established within the Public Works Department Malaysia (PWD). In the same year the Malaysian Government instructed the PWD to prepare a National Slope Master Plan (NSMP) to reduce risks and losses due to landslides. The NSMP was completed and approved by the Government in 2009. There are ten components of the NSMP [1]. They are: 1) Policies and institutional framework; 2) Hazard mapping and assessments; 3) Early warning and real-time monitoring system; 4) Loss assessment; 5) Information collection, interpretation, dissemination, and archiving; 6) Training; 7) Public awareness and education; 8) Loss reduction measures; 9) Emergency preparedness, response and recovery; and 10) Research and development.

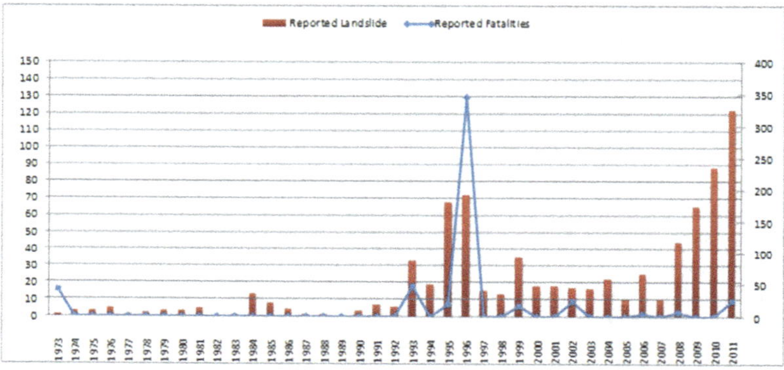

Figure 1: Annual landslides and fatalities in Malaysia.

The NSMP was planned to be implemented in three stages beginning from 2009 up to 2023. Some of the actions taken to manage landslide risks in Malaysia are presented.

2 Data collection, landslide hazard and risk mapping

Lack of a data base on slopes and landslides in Malaysia is one of the reasons why effective decision making on land use planning, maintenance and quantitative assessment of risks has been difficult. Therefore, one of the first actions taken by the Slope Engineering Branch is to collect data by cataloguing slopes for effective slope management. Inadequate information may result in wastage of funding, difficulty in justifying budget allocation, wrong type of treatment for a slope, and difficulty in the preparation of yearly budgets. Even when money is spent on slopes, the safety of most of these slopes may not be assured if information on the slope is inadequate. A hazard rating based on a ranking system known as Slope Management and Risk Tracking or SMART has been developed. The SMART system has been applied elsewhere within Malaysia and is found to be valid for linear application, i.e., road slopes. To date, more than 25,000 man-made slopes along federal roads have been catalogued.

Hazard mapping based on the SMART system and risk mapping of the slopes are in progress.

PWD was also given the task to catalogue slopes and produce hazard and risk maps in high-risk areas, especially cities and towns where sometimes development was carried out without proper planning, design and construction. To date, more than 400 km^2 of hilly areas around Kuala Lumpur, the capital of Malaysia and Penang Island have been catalogued and mapped for hazards and risks. The information obtained from the studies has then been given to the relevant local authorities for development planning, inspection, maintenance, monitoring and other purposes. Several local authorities have used the data to successfully reduce the number of major landslides in their areas. However, not all authorities are proactively using the data given to them to carry out appropriate measures to mitigate landslides. Some of the reasons are due to lack of funds and technical personnel to tackle technical issues.

A budget for more hazard and risk mapping for landslide-prone areas with an approximately area of 1,700 km^2 has been approved. The new project should cover more than 95% of the landslide-prone areas in high-risk areas in Malaysia (as shown in fig. 2).

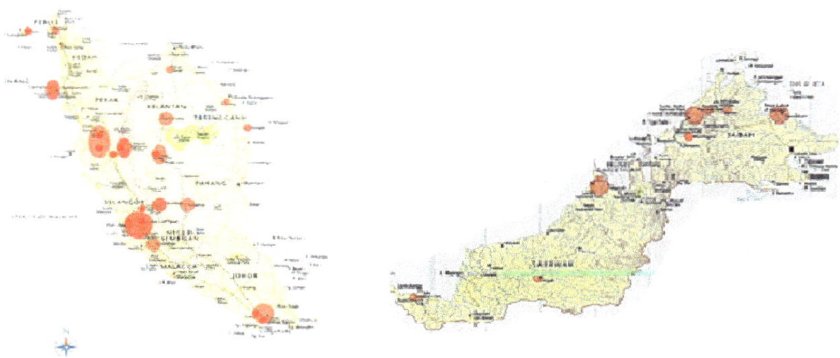

Figure 2: Landslide prone areas in Malaysia are shaded.

The spatial study is divided into two phases, i.e. (i) geomorphological mapping works and (ii) production of hazard map and documentation. In the geomorphological mapping works, Light Detection And Ranging (LiDAR) technology was used in the development of digital terrain models (DTM). Based on this DTM, a ground survey team will carry out field mapping and drainage investigation works. At this stage, qualitative risk analysis was carried out to obtain risk rating for the slope. In the near future, with more data, quantitative analysis will be carried out and risk rating will be given to these slopes.

3 Landslide monitoring and warning system

Landslide warning systems are intended to prompt the public to take precautionary measures to reduce their exposure to risk posed by landslips and to assist engineers, contractors and others who are likely to suffer losses from landslips. The warning also alerts the relevant government departments and organizations to take appropriate actions, such as the opening of temporary shelters, search and rescue operations, closure of schools and relief work [2]. The Landslide Warning (LW) model can be divided into two functions: 1) for monitoring of regions susceptible to landslides; and 2) for critical locations where active landslides are present. For regions susceptible to landslides, a network of rain gauges will be used. It is based on rainfall as a triggering mechanism of landslides. An early warning system based on rainfall pattern and forecasting can be used in an area where studies have been conducted to correlate rainfall with landslides.

For locations where an active landslide is present, instruments such as automatic survey station, piezometer, inclinometer and other sensors can be installed in addition to rain gauges to monitor groundwater level fluctuation and any movement of the unstable slope mass. LW can be issued based on data collected via sensors or instrumentations, warning criteria based on threshold values, communication and information equipment as sensors and human interfaces as tools for dissemination of warning.

For a combination of related matters mentioned above, expertise in various fields is needed in the development of LW. They are listed below:

a. Instrumentation technologies related to landslides;
b. Landslide-related knowledge (process, behaviours, triggering mechanisms);
c. Communication technologies for remote data transmission, and
d. Information technologies for analysis and dissemination of warning.

The output of the LW consists of various levels of warning that can be used by the relevant authorities to execute appropriate measures (such as closing the road or issues evacuation order) and to minimize the effect of the landslides risks if it happens. PWD is discussing with various stakeholders headed by the National Security Council in the Prime Minister's Department to establish a networking system for Landslide Warning System (LWS).

To date, the Slope Engineering Branch, PWD Malaysia has installed more than 30 rain gauges along federal roads. This rain gauge network is necessary infrastructure for LW. The Meteorological Department has also recently installed Doppler radars that have coverage over the whole of Malaysia; however, the radars have yet to be calibrated.

There is an ongoing study carried out by the Slope Engineering Branch for the development of a relationship between rainfall and landslides occurrences to determine the thresholds to be used in the LWS. With an abundance of data on past landslides occurrences over the country and rainfall data obtained from the Department of Irrigation and Drainage and the Meteorological Department, the relationship and thresholds can be determined. A pilot study has started along the Tapah to Cameron Highlands road in Perak. Perak is a state located north of the capital city of Kuala Lumpur.

A slope movement monitoring activity is being carried out at km 44 and 46 along a federal road from Simpang Pulai to Lojing, another road in the state of

Perak. The geology around Mount Pass consists of a sequence of meta-sedimentary rocks that are confined within a 4 km-wide area surrounding the mountain complex. The rocks are highly deformed and adulated, and have undergone low to medium-grade dynamic metamorphism [3]. The original sedimentary rocks are thought to have been deposited during the Ordovician period [4]. During the late Palaeozoic period it was deformed and metamorphosed. The meta-sedimentary rocks have been intruded by granite plutons that are predominantly Permian to Jurassic in age [5]. The exposed rock cut slope within the site includes quartz mica schist, graphitic schist, quartzite and phyllite with weathering grades varying from Grade II (slightly weathered) to Grade VI (residual soils). The residual soils are confined to the 3 to 6 m top section of the slopes. According to a study by Andrew Malone Ltd [3], the rock sequence is cut by a series of faults that are grouped into 3 categories. The first is the most prevalent within the graphitic schist unit, consisting of low to moderately dipping faults that are associated with quartz veining. The second consists of steep faults that cut sharply across the foliation.

At km 44, a cut slope has translated downward and outward by more than 30 m. Even during construction, this slope had already started to fail. It was initially designed for a cut of no more than 40 m high, which is quite common in Malaysia when tunnels and viaducts are deemed too expensive compared with cut and fill. PWD has drafted the Slope Design Guidelines [6] to limit the height of cut and fill slopes to less than 40 m to facilitate maintenance of the slope and reduce the environmental impact of the construction. At the same time the guidelines encourage the construction of tunnels and viaducts, which is more environmentally friendly. This case clearly illustrates that the present method of cut and fill may not necessarily be more cost effective compared with tunnels and viaducts in the long run. Monitoring of the slope movement is carried out in real-time using a surface monitoring device to measure the movements of the

Figure 3: Landslip at Gunung Pass and location of one of the prism markers.

slope surface. In this case, two automatic total stations with prisms as markers are used to monitor the movements of the slope. Some of the actions taken include conducting round-the-clock inspections, monitoring and minor realignment of the road. Fig. 3 shows the head scarp at the top of the mountain and the location of the markers on the slope surface.

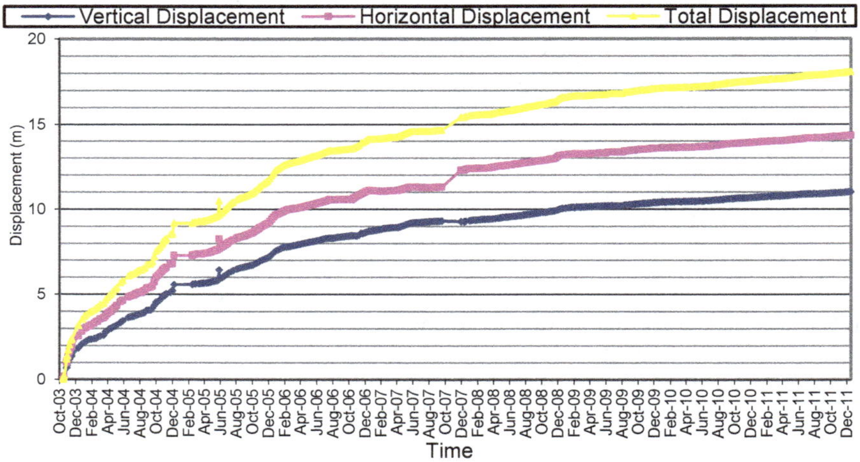

Figure 4: Largest movements at prism marker indicated in fig. 3.

Movements of the slope surface have been monitored since 2003. Initially, the monitoring was done manually until 2007 when two automatic total stations were installed and data was acquired at three-hour interval as opposed to once a day when measurements were taken manually. Fig. 4 presents the movements of the most active marker near the head scarp. The slope face is still moving at a rate of 0.4 m per year.

4 Slope inspection program

Slope inspection for all federal roads is carried out in preparation for the rainy monsoon seasons by the PWD technical staff with the help of the district engineer's staff. A few local authorities are also carrying out regular inspections on the slope under their authority. Efforts have been made to create awareness amongst the local authorities so that regular inspection and maintenance are carried out to reduce the risk of landslides.

Emphasis during the slope inspection is placed on ensuring that the drainage system is well maintained, i.e., free of debris and any damage has to be reported and repaired. Another aspect of slope inspection is to detect signs that may indicate or cause instability such as tension cracks and seepage within the slopes. This program has just been initiated and guidelines for maintenance have been introduced. At the outset the process seemed quite simple. However, ensuring the effectiveness of the program requires a lot of groundwork, such as training for the inspectors – usually technicians and engineer's assistants, proper equipment and, of course, enough personnel to carry out the inspections. Other impediments to the slope inspection program include inadequate facilities for inspection purposes. Usually cut slopes in Malaysia are constructed at more than 35° and some rock slopes are constructed at more than 75°, which means that a special ladder or stairs have to be provided for scaling the slopes. In actual fact, such facilities are generally ignored during the design. On top of this problem, the slopes are generally heavily vegetated. PWD is trying to remedy the situation by constructing facilities to facilitate inspection of the slopes.

Previously, there was no limit to the height of the cut slope, so the highest road cut slope is approximately 260 m. Many slopes are more than 40 m high which means that only very physically fit personnel would be able to scale the slopes. Coupled with the inadequate facilities as mentioned above, the task of proper inspection is difficult. Apart from manual inspection, PWD are looking at ways to carry out remote inspection. One of the ways is to inspect using unmanned model helicopter or plane with viewing device attached; however, this is still undergoing trials and experiments.

Once the inspection process is completed, the next step would be to perform proper maintenance and repair. Apart from the physical impediments mentioned above that makes maintenance and repair difficult and costly, adequate funding is always an issue. For routine maintenance and minor repairs of road slopes, PWD is provided with approximately USD 33 million annually.

5 Guidelines

Manuals and guidelines facilitate routine works and procedures that need to be done quickly and efficiently. Several guidelines have been produced to help engineers, especially those in PWD to carry out their works. Guidelines are sometimes adopted with some modification from other established organizations that have already come up with a proper procedure.

Some of the guidelines that have been produced are slope design and maintenance guidelines. In the slope design guidelines, some cut slope and embankment dimensions are clearly spelled out, for example the maximum number of benches are now limited to 6 and the height between each bench shall not exceed 6 m. The minimum bench width of the slope is also specified. The factor of safety cut slope and treated slopes are specified.

Malaysia's slope maintenance guidelines have been adopted with some modification from the Hong Kong slope maintenance guidelines. In the guidelines, inspection program for slopes is presented including the personnel and the frequency of inspection. The guidelines also include what needs to be maintained and how it should be carried out. As in any other maintenance guidelines, special attention is paid to the condition of drains on the slopes. Any signs of instability have to be attended speedily before the condition becomes worse. Examples of well and poorly maintained slopes were shown by way of pictures and illustrations. The maintenance manual can be downloaded for free from PWD's website and the maintenance guidelines have been distributed to local authorities for maintenance of their slopes, which usually include slopes along municipal roads and around residential areas.

The Town and Country Planning Department, a federal agency, has also drafted development on hilly terrain and highlands guidelines in 2009 to ensure man-made slopes are well planned, suitably designed, well-constructed and properly maintained. The guidelines are mandatory for developers if the local authorities choose to enforce them. Some states have since drafted more stringent guidelines to reduce the risks of landslides. Public pressure has resulted in the federal and state government to take more proactive actions to reduce major landslide disasters.

Work on a regional rainfall-induced landslide warning system has started with some research being carried out along the road between Tapah to Cameron Highlands, a road that traverse through a hilly terrain area.

6 Training

In this component, two strategies were proposed to increase knowledge and awareness among students, practicing engineers, stakeholders such as developers, local government officers, contractors, planners and geologists. In the case of students, one of the recommendations was to introduce curriculum and course templates for undergraduates and post graduates courses. Discussions with the academia from various local universities during the drafting of the NSMP did not yield a favourable response from them simply because there is no space for a new course to be taught within the already congested course schedule. It was suggested that special talks be given by practising engineers on relevant topics on slope engineering to under and post graduates students. The programme has already started with a few universities and will be expanded to other institutes that teach civil engineering and soil sciences [7].

Currently, more effort is spent on training engineers and technical personnel from the local authorities so that they can respond to the demand from the public to provide technical advice, carry out inspection on slopes and recognising signs of landslides so that they can take preventive actions should the need arises.

Other stakeholders include practicing engineers, planners, developers, contractors and geologists who would require training and awareness programmes. Developers and contractors in particular would require such training so that their works have better quality. PWD plans to work with the Malaysian Construction Industry Development Board to carry out a systematic programme of courses and seminars for the construction industry players on good practices on slopes.

7 Public awareness and education

One of the first programmes to be rolled out upon the completion of the NSMP was public awareness and education. The programme was about creating awareness of slope safety by minimising the effects of landslides through proactive actions and measures that can be taken by community members as well as by government and private owners of slopes.

Generally, people tend to focus on safety only after a disaster happens, so the programme aimed at getting them thinking about averting disasters before they occur [8].

Although the main target groups of the programmes were the communities-at-risk and the general public, there were other target groups consisting of the state and local governments, private slope owners, media, universities and schools.

The objective of the awareness programmes was to convey two key messages to the public. The first was to let the public know that there is a body of useful information that is available to the public on the phenomenon of landslides and tips on monitoring and maintenance. The second is that there is a government agency dedicated to safeguarding the interest of public safety.

Figure 5: Children lining up to receive prizes in a colouring contest in one of the public awareness programmes.

These messages were encapsulated in the campaign theme of "Learn, Maintain, Monitor and Report" and all activities of the awareness programmes were located around this theme. The motif that tied all these activities together was the slogan "Safe Slopes Save Lives", courtesy of the Geotechnical Engineering Office in Hong Kong.

One of the most important groups is the communities in at-risk areas because of the obvious risk to life and property. The assumption at the outset of the programme was that public awareness to this group would yield the best results among all the target groups because of the immediate safety concern to themselves.

Next to communities-at-risk, the most important target group was the local authorities. The authorities are the only government body with the charter to enforce safety guidelines and by-laws and engage in maintenance measures. Because of their being the first line of contact with the residents, it is crucial that the engineering departments of the authorities are well-trained and well-equipped.

The outcomes resulting from this programme are significant in that they reflect institutional and long-term changes that affect the way hillside developments will be carried out in the future. They are as follows:

1. Awareness among local authorities in hilly areas for a proper slope management mechanism within their scope of work and the subsequent establishment of Slope unit within 3 local authorities in the high-risk states in the country
2. Establishment of state-level independent slope oversight committees for checking and approving all new development orders involving hills.
3. Reporters of newspapers providing regular coverage of slope issues in the local beat, and coverage now includes educational material in addition to problem cases.
4. Formation of SlopeWatch, a community-based organization that has grown into an non-governmental organization due to demand by residents for more information and advice on averting slope problems and pushing the authorities for stricter supervision of developers.
5. Residents associations in urban areas establishing sub-committees on slope monitoring so that residents can do their own monitoring and report to the authorities on any signs [8].

These are some of the changes that the programme has affected.

8 Inter-governmental Committee on Slope Management

One of the action plans in the National Slope Master Plan is the setting up of the Inter-governmental Committee on Slope Management (ICSM) [9] at federal and state levels. Federal-level ICSM was set up in 2010. The primary purpose is to provide a mechanism for the active participation of all stakeholders, including technical experts and communities, in slope management planning and operations, where relevant role players consult one another and coordinate their activities on slope management issues.

The ICSM members include but are not limited to the following organisations: Slope Engineering Branch; National Security Council; Town and Country Planning Department; Department of Environment; Department of Mineral and Geoscience; Malaysian Meteorological Department; Malaysian Centre for Remote Sensing; PWD Malaysia; Malaysian Highway Authority; Department of Survey and Mapping; Royal Malaysian Police Force; Malaysian Fire and Rescue Department; Social Welfare Department; the Red Crescent; highway concessionaires and other relevant governmental and non-governmental organisations.

The ICSM must advise and make recommendations on issues relating to slope management and the establishment of the landslide disaster management framework. The committee has recently agreed to form sub-committees headed by National Security Council to review the efficiency of the rescue operation for landslides and to facilitate information exchange among various agencies not only on landslide matters but also other information that any of the agencies have acquired.

9 Conclusion

Some of the slope management issues and actions taken by the Slope Engineering Branch of PWD Malaysia and other agencies are presented. PWD is at the forefront of managing landslide risks in Malaysia. The process commenced as soon as the Slope Engineering Branch was established within PWD and the drafting of the NSMP. The Malaysian Government has started to pay more attention to the landslide issues as public outcries grew louder. Some local authorities that are at the front line of public anger whenever major landslides occurred have yet to provide better services while others have used guidelines and relevant acts to act against errant developers and land owners. Landslide risk management requires the cooperation from all sectors of the society, from the federal and state government, local authorities, non-governmental organisations and the public for its success.

References

[1] Public Works Department, *National Slope Master Plan*, Slope Engineering Branch, PWD Malaysia, pp. 1-2, 2009.

[2] Hong Kong Observatory, Landslip warning. http://www.hko.gov.hk/wservice/warning/landslip.htm

[3] Andrew Malone Ltd. *Final report*, Landslide study at Ch 23+800 to Ch 24+460, Simpang Pulai – Lojing Highway, 2006.

[4] Geological Survey of Malaysia, Geological Map of Peninsular Malaysia, 8th Edition, 1985.

[5] Gobbett, D.J. & Hutchison, C.S., *Geology of Malay Peninsula, The Geological Society of Malaysia,* Wiley-International: New York, 1973.

[6] Slope Engineering Branch, *Slope Design Guidelines*, Public Works Department of Malaysia, 2010.

[7] Abdullah, C.H., Malaysia's national slope master plan – from theory to practice. *Proc. of the 11th International and 2nd North American Symposium on Landslides and Engineering Slopes*, eds. E. Eberhardt, C. Froese, A.K. Turner & S. Leroueil, CRC Press: AK Leiden, pp. 135-152, 2012.

[8] Motoyama, E. & Abdullah, C.H., Landslide public awareness and education programs in Malaysia. *Proc. of the Second World Landslide Forum,* eds. F. Catani, C. Margottini, A. Trigila & C. Iadanza, CSR: Rome, Abstract p. 544, 2011.

[9] Public Works Department, *National Slope Master Plan*, Slope Engineering Branch, PWD Malaysia, pp. 5-5, 2009.

Early warning about regional landslides using rainfall characteristics

J. Oh, Y. Son, Y. Park, Y. Song & M. Jung
Ministry of Public Safety and Security, Korea

Abstract

In order to reduce damage by landslides, an early warning system can be one of the effective countermeasures based on the early detection of landslide occurrence and, therefore, lead to pre-emptive early evacuation of residents from a landslide prone area. Early detection of landslide indication in a broad natural terrain can be performed by monitoring rainfall in real time or in near-real time. Based on the real time rainfall data, this study suggested a method to warn of landslides by the application of a tracking line of rainfall. The method makes it possible to set up warning criteria for landslides considering local rainfall characteristics in the summer. This study suggested a method of landslide early warning based on the regional rainfall conditions.
Keywords: landslide, rainfall threshold, early warning system, slope disaster.

1 Introduction

Every year, Korea suffers from various seasonal natural hazards, which cause tremendous property damage and human losses. Major types of natural hazards in Korea are typhoons and heavy rains, but an earthquake is rare. Another natural hazard, drought, causes adverse impacts on agricultural lands, but the actual damage can be hard to measure. Korean economic and social development in the last few decades has led to a number of populated and quickly industrialized cities. Modernized, industrialized and urbanized Korean cities show a somewhat different trend from previous cases in terms of disaster characteristics.

A variety of disasters are known and large-scale disasters are reported that are not influenced by time and national or regional borders because of urbanization and densification in modern society. For example, the collapse of the Samphoong department store in 1995 and the impact of Typhoon Rusa in 2002 well illustrate the changing face of disasters in Korea. The Indian Ocean tsunami disaster in Asia and Cyclone Nargis in Myanmar are recent warnings for disaster managers. In the future, vulnerability will be increased because of difficulties in risk prediction and efficient response, the aging effect of infrastructures and the probable increase of new societal disturbance such as failure of information and communication systems [1].

In Korea, human and property damage occur every year due to geological disasters, especially, slope failure-related disasters that are the result of heavy rainfall due to the typhoon and monsoon from July to September. Types of slope-related disasters include cut-slope failure along the roads, natural terrain failure in mountain areas and retaining structure failures in the residential areas. The human vulnerability resulting from these geological hazards has been increasing as a function of industrialization and population growth.

Countermeasures to these disasters, however, are limited to field surveys after disasters in Korea. Also, most measures are concentrated on emergency recovery and hence insufficient effort is given to prevent secondary disasters. Therefore, it is necessary to improve the disaster management system for the geological hazards, not only concerning the engineering part, but also with regard to the political and institutional areas.

2 Geotechnical disasters in Korea

The vulnerability against landslides is surely increasing because of landscape change and sometimes badly engineered cut slopes. As shown in table 1 (modified after Park *et al.* [2]), human casualties due to various natural disasters are not insignificant. The most serious natural disaster in Korea is flash flood

Table 1: Human loss due to various natural disasters in Korea (modified after Park *et al.* [2]).

Year	Death by various natural disasters	Death by steep slope disasters	Ratio (%)
1998	384	111	28.9
1999	89	32	36.0
2000	49	12	24.5
2001	82	9	11.0
2002	270	79	29.3
2003	148	37	25.0
2004	14	3	21.4
2005	52	11	21.2
2006	63	22	34.9
2007	23	4	17.4
2008	16	8	50.0
2009	12	6	50.0
2010	14	1	7.0
2011	78	57	73.1
Total	1,306	392	30.0

followed by landslide-related disasters. In the last 14 years from 1998 to 2011, disasters in steep slope areas killed 392 people.

Annual average of people killed by slope-stability related disasters such as landslides and debris flows for the last 14 years is 28 in Korea. Even though many technologies are developed for landslide mitigation, the extent of damage by landslides is not decreasing.

One of the reasons for this is the reckless developments near housing areas and the absence of regulations or enforcement to consider slope stability problems in the area and no warning systems for the residents near hillsides in Korea. Figs 1 and 2 show typical slope-stability related disasters in cut slope and natural terrain, respectively.

Figure 1: Destroyed residential area and beltway by natural slope failure.

Figure 2: Buried houses by debris flow in Gangwon Province in 2006.

3 Suggestion of landslide early warning methods based on rainfall characteristics

The early warning system for landslide disasters using rainfall data in this study is based on an empirical approach. It assumes that an area which experienced landslides with specific rainfall characteristics will experience similar landslides when similar rainfalls occur. If rainfall characteristics in a specific area can be observed in real time, an early warning for that specific area may be possible.

Therefore, before developing an early warning system, it is necessary to compile available information on previous landslide history with exact occurrence time and rainfall data. Previous rainfall accumulation data are analyzed and manipulated to find several rainfall characteristics, such as rainfall intensity and antecedent rainfall, as well as the relationship between rainfall and landslide occurrence. There are several steps or processes (fig. 3) in a landslide early warning system before issuing landslide warning using various means of information dissemination to the local people such as sirens, automated phone calls and mass media broadcasting.

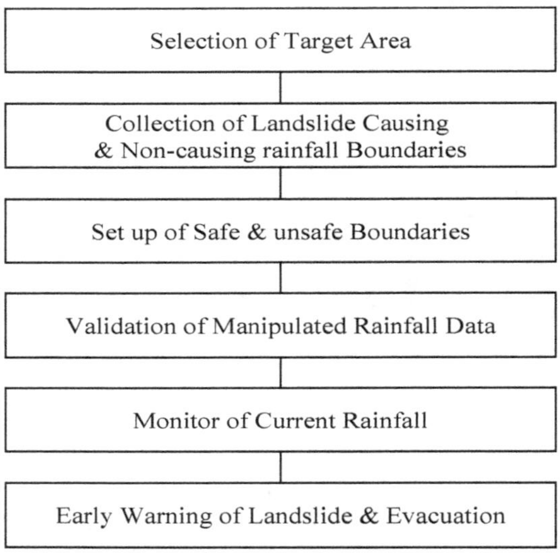

Figure 3: Flow of landslide early warning.

The first step is to find landslide-causing and landslide-non-causing rainfall data. This process is critical since the accuracy of early warning system is highly dependent on these data. Many data points can be depicted depending on the rainfall and landslide data availability. To consider the decaying effect of antecedent rainfall, half-life is applied depending on the site conditions. The half-life is the time period over which the rainfall amount in the soil or on the ground surface reduces to half reflecting the effect of rainfall dissipation. Most of the previous research that adopted rainfall intensity and accumulative rainfall for the relationship between landslide occurrence and rainfall characteristics did not consider the effect of rainfall dissipation in the soil. The accumulative rainfall with decaying effect of antecedent rainfall can be defined as a working rainfall R_w as expressed by

$$R_W = \sum \alpha_i \times R_i \quad \text{and} \quad \alpha_i = 0.5^{i/T} \tag{1}$$

where α_i is the decay coefficient at i hours before a landslide; R_i is rainfall intensity at i hours before a landslide; and T is the half-life. The decay coefficient was estimated by the tank model, which is one of the rainfall runoff models; the first layer and third layer in the tank model represent surface flow and ground infiltration, respectively [3].

In this study, to consider antecedent accumulative rainfall amount and rainfall intensity (as shown in fig. 4), working rainfalls in abscissa and ordinate were calculated based on half-lives of 72 hours and 1.5 hours, respectively, proposed by the Ministry of Land Infrastructure and Transport (MLIT) [4]. When the decay coefficient becomes less than 0.004, the effect of antecedent rainfall is minimal. When applying 1.5 and 72 hour half-lives, the coefficients at 12 and 574 hours (about 24 days) before a landslide, respectively, become less than 0.004. To estimate more precise half-life representing soil characteristics in Korea, studies on the conversion of infiltration characteristics into applicable half-life will be needed based the relationship between rainfall and infiltration.

The second step is to draw a boundary or the critical line between rainfalls that caused and did not cause landslides. Generally, the upper right side in the graphs of fig. 4 is an unsafe zone, i.e. rainfall data which caused landslides are depicted. In this study landslide causing and non-causing rainfalls were surveyed in the field and analyzed using data from an automatic warning system (AWS) located in the test site (fig. 5). A straight line was selected between causing and non-causing area and other shapes for the critical boundary can be considered.

The third step is to set up warning and evacuation lines in the same graphs. Those lines are needed to be drawn considering enough evacuation time at a specific village. Enough evacuation time should reflect local characteristics such as how far the shelters are located, how many vulnerable people live in the town, and how fast the disaster information can be disseminated to the people after the warning. Those lines are usually predetermined, as in this study, based on a maximum of half-hour and maximum one-hour rainfall, respectively. After establishing these lines and boundaries in a target area, the generated graph should be checked whether it is truly representing rainfall characteristics. If precise rainfall prediction is possible using radar or other instruments and the exact time when predicted rainfall will penetrate the critical boundary can be estimated, those lines do not need to be predetermined.

The fourth step is to monitor current rainfall in a specific area where the graph is prepared for and analyze rainfall data drawing a tracking line. When the tracking line crosses the warning or evacuation line, an early warning and evacuation order can be disseminated. The final step in the landslide early warning system is to make actual warning for the people and make them evacuate from dangerous landslide areas using various means of devices and laws and regulations.

In this study, tracking lines are drawn and monitored based on those real-time data according to the various preset lines such as critical lines, evacuation lines, and warning lines, which were selected considering maximum rainfalls in one and two hours, or maximum rainfalls in 30 minutes and one hour in the area. The tracking lines did not reach the warning lines which have been set up based on

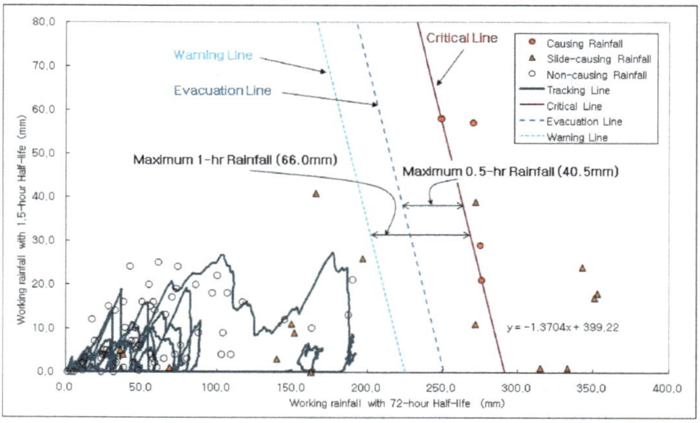

(a) Garisan elementary school area

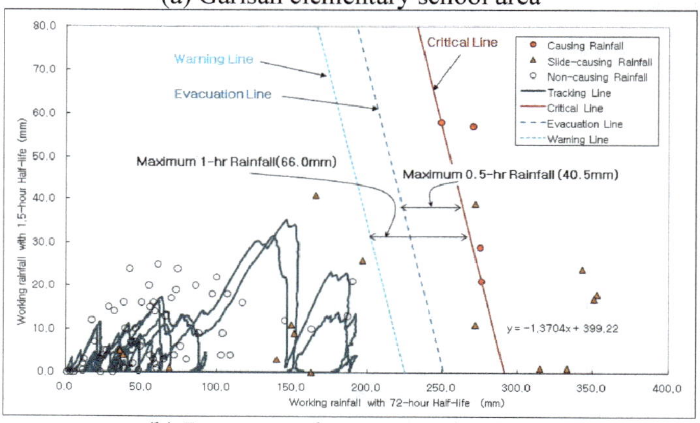

(b) Bupyeong elementary school area

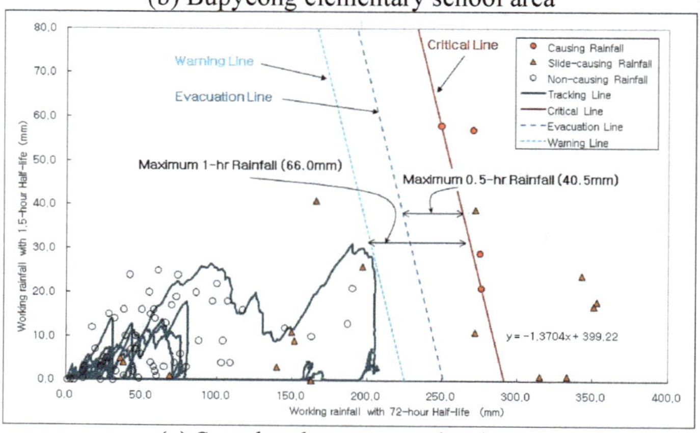

(c) Seowha elementary school area

Figure 4: Application of tracking line by the real-time data.

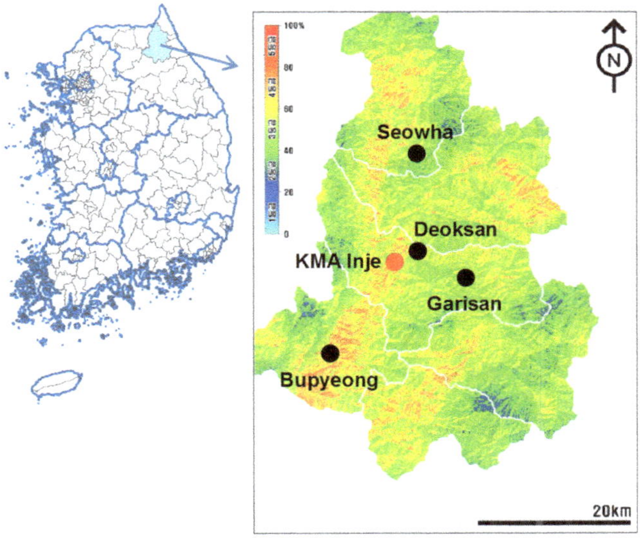

Figure 5: Location of rainfall monitoring system in the study area.

the maximum rainfalls in half hour and one hour. Also, all the tracking lines did not reach the critical lines regardless of the selection of the maximum rainfalls during the summer of 2008. This finding can be considered as evidence to support the validity of this study's approach, since there were no reported steep slope failures or landslide disasters in the study area during the summer of 2008 [5]. The application of tracking line generated by the real-time data from Garisan, Bupyeong and Seowha elementary schools are shown in fig. 4 using maximum rainfalls in half an hour and one hour for evacuation and warning lines, respectively. The filled circles, empty circles and triangles in the figure represent, respectively, rainfalls that actually caused human casualties, rainfall events with significant rainfall amounts without landslide occurrences and rainfalls that caused landslides but with no human casualties. Critical lines, which are the same for the three sites in the study area, are drawn at the boundaries of rainfalls that caused human casualties by landslides in the study area. Tracking lines in the figure, which are different from each other in the three sites of the study area, show several rainfall events in the study area during the summer of 2008. Since the decaying effects of rainfalls are reflected in the tracking lines, the lines can converge to zero point when a rainfall reduces or stops. This attribute of tracking lines can be applied when cancelling or calling off warning and evacuation orders.

4 Conclusions

Steep slope related disasters such as landslides occur almost every year in Korea causing social and economic disturbances. Fundamental countermeasures that can prevent steep slope related disasters are not always easy to implement due to uncertainty of disaster location and budget limitation. For an effective preparedness and response, a landslide early warning system can be considered as part of preemptive measures to minimize human casualties. Early detection of landslide indication in a broad natural terrain can be performed by monitoring of rainfall in real time or in near-real time.

In this study, three rainfall monitoring gauges have been installed in three different elementary schools in Seowha, Bupyeong, and Garisan to suggest warning criteria for landslides considered with local rainfall characteristics. The difference of rainfall data clearly explains that local rainfall patterns are strongly dependent upon topography, especially upon elevation in a mountain area as well as distance of each observation position during the summer. Therefore, it can be articulated that the density of AWS needs to be increased to be used as local standard for rainfall measurement unless rainfall measuring systems are installed independently for the operation of landslide early warning system. Tracking lines were drawn and monitored based on those real time data according to the various preset lines such as critical lines, evacuation lines and warning lines, which were selected considering maximum rainfalls in one and two hours, or maximum rainfalls in half hour and one hour in the area. The tracking lines did not reach the warning lines which have been set up based on the maximum rainfalls in half an hour and one hour. Also, all the tracking lines did not reach the critical lines regardless of the selection of the maximum rainfalls during the summer of 2008. This finding can be considered as evidence to support the validity of this study's approach, since there were no reported steep slope failures or landslide disasters in the study area during the summer of 2008.

References

[1] Park, D., Oh, J., Park, J. & Chae, B., Slope stability related disasters and regulatory countermeasures in the Republic of Korea. *IPL 2007 Symposium*, Tokyo, Japan, pp. 19-23, 2007.

[2] Park, D., Oh, J., Son, Y., Park, J. & Lee, M., Steep slope management system in Korea. *East Asia Landslides Symposium*, Seoul, Korea, pp. 7-11, 2008.

[3] Sugawara, M., Ozaki, E., Watanabe, I. & Katsuyama, Y., Tank model and its application to Bird Creek, Wollombi Brook, Bikin River, Sanaga River and Nam Mune, *Research Notes*, National Research Center for Disaster Prevention, No. 11: pp. 1-64, 1974.

[4] MLIT, *Guidelines for Development of Warning and Evacuation System against Sediment Disaster in Developing Countries*, Infrastructure Development Institute, Ministry of Land Infrastructure and Transport, p. 116, 2004.

[5] NIDP, *Study on the Applicability of an Early Warning system using Rainfall Data in Korea*, Research Report, National Institute for Disaster Prevention, National Emergency Management Agency, p. 320 (in Korean), 2008.

On spatial uncertainty in hazard and risk assessment

A. G. Fabbri[1] & C.-J. Chung[2]
[1]*DISAT, University of Milano-Bicocca, Italy*
[2]*SpatialModels Inc., Canada*

Abstract

Exploiting the mathematical framework of favourability function modelling and of software designed for spatial target mapping, experiments are discussed to measure the effects of uncertainty of spatial support. A database that has been the focus of landslide hazard prediction and risk assessment of buildings, roads and land uses is reanalyzed gradually modifying the spatial characteristics of supporting patterns of the proposition of "*presence of landslide occurrences.*" Boundary fuzziness of categorical mapping units and filtering of continuous fields, represent weakening of spatial relationships. Comparisons are made of ranges of ranks representing uncertainties of class membership in target patterns obtained by iterative cross-validations. The impacts of the different spatial uncertainties on risk assessment patterns are visualized using progressive combinations of uncertainties of target patterns to resolve risk equations and study the consequent changes in risk values.
Keywords: favourability modelling, target mapping, spatial uncertainty, spatial support, landslide hazard, landslide risk.

1 Introduction

It is now over fifty years that attention is being given to spatially distributed data for its integration into indices for development planning. Examples are resource exploration of mineral occurrences or locating hazardous sites such as zones likely to be affected by landslides or by floods. Most emphasis has been on methods of spatial statistics and on the introduction of expert's opinions to complement data scarcity or insufficiency.

The abundance of digital data today and the consequent feasibility of constructing well focused spatial databases for regional planning provide further challenges: testing and interpreting the information content of the databases and navigating through the many options encountered to generate manageable representations. Instances of these are the spatiotemporal distribution of future discoveries or that of future hazardous occurrences. It means that not only a database must adequately and acceptably represent the distribution and dynamics of natural processes but also should allow measures of spatial uncertainty. In practice it must offer criteria for deciding on the relative significance and strength of the spatial support. The task becomes similar to the one in computer war games in military exercises where sequences of simulations are generated to predict consequences of modifying spatial relationships.

In this contribution the impacts of different spatial uncertainties on landslide risk assessment patterns are visualized using progressive combinations of uncertainties of target patterns to resolve risk equations and study the consequent changes in risk value distributions. A database that has been the focus of landslide hazard prediction and risk assessment of buildings, roads and land uses is reanalyzed gradually modifying the spatial characteristics of supporting patterns of the proposition of "*presence of landslide occurrences*." Following a description of the database, a brief introduction is provided to favourability modelling and its application software. Experiments are then discussed on landslide hazard prediction and risk assessment in the Deba Valley in northern Spain. The spatial data have been modified to weaken the spatial support and degrade the quality of prediction and assessment introducing levels of greater uncertainty. Conclusions are drawn on the importance of simulating spatial changes.

2 The Deba valley spatial database

The Deba Valley is part of the Basque Province of Guipuzcoa up from the coast of northern Spain. The study area is located in the lower part of the valley and covers approximately 140 km^2, with maximum elevation just below 700 m a.s.l. Main annual rainfall reaches 1500 mm with episodes of over 100 mm/day every few years. Lithologies in the area are: limestone, marl, clay stone, sandstone, flysch and volcanics, of the Cretaceous and Palaeogene of the Basque-Cantabrian Pyrenees. The average slope gradient is about 22° and regolith thickness ranges from 50 cm to 3 m. Shallow translational landslides and flows triggered by rainfalls are the most common types of mass movements in the area whose landscape is highly influenced by reforestation, cultivation, urbanizations, infrastructures and industrial activities. Population density reaches 500/km^2.

Remondo *et al.* [1–3], have constructed a digital database for landslide hazard studies later extended for risk assessment. Through photo-interpretation and field work, 1123 shallow translational landslides and associated flows were mapped and dated: 906 prior to 1997, and 217 for the period 1998–2001. The average size of their trigger zones is approximately 400 m^2, so that it was decided to represent each by a single picture element or pixel of 10 m resolution. The same digital resolution was used for the rest of the database that was to represent the typical setting of the landslides: 25 lithologic units, 9 land use classes (both categorical), and elevation, aspect, curvature and slope (continuous fields). The study area falls within a rectangular raster of 1886 pixels by 1555 lines and occupies 1,393,541 pixels.

Remondo *et al.* [2] first modelled landslide hazard in the area to predict the distribution of future likely landslides using the spatiotemporal distribution of the 1123 landslide trigger zones and their spatial relationships with the categorical and continuous digital maps. Later, Remondo *et al.* [3] obtained a semi-quantitative risk-assessment augmenting the database with socioeconomic spatial elements: 5 types of roads, 3194 buildings and 9 types of land uses. They were made into co-registered digital maps for which values and vulnerabilities were estimated and compiled as 10 m pixels.

The Deba Valley spatial database that is used in this contribution has also been turned into a case study for training decision makers [4].

3 Favourability modelling: approach, software and strategies

Favourability modelling is based on constructing a proposition within the study area as a mathematical statement that has to be proven true or false given the spatial evidence. Such evidence is established for the study area as a function of the spatial relationships established between a direct supporting pattern (DSP) of the proposition and indirect supporting patterns (ISPs). In the Deba Valley study area, for instance, the DSP is the spatial distribution of a well defined set of occurrences, such as the 1123 shallow translational landslides and associated flows. The ISPs are then the distributions of the various categorical mapping units and the values of the continuous fields used to represent the typical settings of the DSP: lithologies, land use classes and values of continuous fields of topographic aspect, curvature, digital elevation and slopes.

The spatial relationships are conveniently calculated from co-registered digital images with a given common spatial resolution: in our case of 10 m for all DSP and ISPs. Example of a proposition is: "P_i*: a point i in the study area is affected by a part of a future landslide of type shallow translational landslide and associated flows | given the presence of the classes and values of spatial evidence.*" Different interpretations are possible for such a favourability function: possibility, likelihood, certainty, belief and plausibility, conditional probability and more, using the corresponding modelling assumptions. Here our analyses will be limited to the empirical likelihood ratio (ELR) function, amply discussed elsewhere [5]. The categorical ISPs are transformed into normalized frequencies and the continuous fields into density functions, so that the ELR ratios are computed between the functional values in the presence of the landslide trigger zones with those in their absence. Ratios for overlapping ISPs are combined by means of the rules and assumptions of the model. ELR values range from 0 to ∞.

The application of a favourability function must be properly structured in time and space and can be used as predictor of future landslide occurrences via assumptions or scenarios of spatiotemporal nature, e.g., similarity of settings through the study area, of database sufficiency, or similarity of frequency of occurrence through time etc.

Criticism on applications of favourability modelling [6] apparently justified the programming of STM, a software for spatial target mapping developed for research and training [4, 7] as a tool accessory to conventional geographic information systems (GIS). Besides data transfer to and from GIS, STM provides several prediction models: fuzzy sets, likelihood ratio, linear and logic regression, and Bayesian probability functions. In addition, it allows: spatial data input modifications, modes of iterative cross-validations, generation of prediction, target, uncertainty and uncertainty/target combination patterns, with associated statistics and prediction-rate tables. These will be described in the application section that follows.

In particular, with the selection of ISPs as input to modelling, transformation parameters are available. Linked with the digital maps of categorical ISPs, such as lithology and land use, inputs are the respective boundary images of the units as thin lines of one-pixel width. A *number-of-pixels-in-neighbourhood* parameter allows the generation of fuzzy boundaries of desired width, say 5, 10, 15 etc. For the continuous fields ISPs, a *spread parameter* of continuous values is used to obtain an appropriately smooth distribution before proceeding with further analyses. These two parameters are indicated as N and S, respectively, used for artificially fuzzyfying categorical boundaries and smoothing continuous fields.

Of a study area, STM generates equal-area rankings as classifications termed prediction patterns. The statistics for the DSP/ISPs spatial relationships can also be extended from a study area to another, assumed to have similar settings and to generate a prediction pattern there. Within a study area, the DSP can be partitioned to obtain sequences of prediction patterns and of associated statistics by iterative cross-validations according to one of the following strategies:

(a) Sequential selection of a given number of occurrences as DSP;
(b) Sequential exclusion of a given number of occurrences as DSP;
(c) Random selection of a given number of occurrences as DSP.

By cross-validation, the statistics of the relative distribution of the occurrences in the equal area ranked classes of the prediction pattern is obtained as a prediction-rate table. The selected subset of the available occurrences is used to apply the modelling and the remainder is used to study their distribution across the equal area ranked classes obtained by modelling. From the iterative processing strategies in (a), (b) or (c), or other combinations of those, a sequence of prediction patterns is generated. A target pattern combines them by means of a number of optional statistics: basic statistics such as sample average and variance or sample median and range, jack-knife statistics as average and range, or rank-based statistics as median range and range of ranks.

For instance, a target pattern can be a digital image with pixel values as means of a sequence of prediction patterns from iterations in (a), (b) or (c). An associated uncertainty pattern of class membership in the target pattern has ranked values that represent the standard deviation at each pixel. Furthermore, a given % of the lower rank values, corresponding to a narrower range in an uncertainty pattern, can be used to select only the values in a target pattern that correspond to those lower uncertainty values, thus obtaining a combination pattern of uncertainty and target.

Iterative cross-validation strategies allow establishing the relative reliability, stability or acceptability of the prediction and target patterns. These are represented as prediction-rate tables, histograms and cumulative curves.

The associated SRA software for spatial risk analysis [6, 7] is complementary to STM in that it takes as part of the inputs a prediction pattern and a prediction-rate table. The table contains the cumulative frequency of occurrences for all equal area ranked classes of predicted hazard and is then converted, via a scenario, into a probability of occurrence histogram either directly or via a monotonically decreasing transformation or as a fitted function.

In addition, SRA requires as inputs co-registered categorical images of socioeconomic indicators representing elements exposed to risk, such as roads, buildings, land uses (and persons in case of casualties), with accompanying tables of vulnerabilities and estimated values. From all those inputs, SRA calculates Risk Assessment patterns and associated statistics, either for single element type or as aggregation of them.

The following section deals with experiments that exploit many of the features of STM and SRA described here. The purpose is to expose aspects that contain more general significance out of the Deba Valley database.

4 Weakening of spatial support

Experiments on landslide hazard prediction and risk assessment in the Deba Valley study area have been discussed in a number of studies [2–4]. The aim here is to explore situations in which the support for spatial relationships becomes gradually weaker to simulate poor quality of prediction/risk patterns. This is to provide insight on the question: “when to decide that a database is too poorly representing future hazardous occurrence distribution?” Or, in other words: “how to compare a poor prediction with a good one?”

The application performed here uses the digital image with the one-pixel distribution of the 906 pre-’97 landslides and that of the 217 post-’97 as DSP that are shown in figs 1(a) and 1(b), respectively, with the study area as background. As ISPs to characterize the settings of the landslides, it uses two categorical digital images, lithology and land use, termed *l* and *u*, and four continuous ones, aspect, curvature, digital elevation and slope, not shown here, termed *A, C, D* and *S*. The ELR model is first applied using the entirety of 1123 landslides as DSP and the *luACDS* as ISPs. This is because the prediction pattern obtained from them is the most detailed as it uses all the DSP data available. However, the relative quality of its classes is not known as yet without performing their cross-validation. This is done by repeating the modelling using only the 906 pre-’97 landslides and verifying the ranking of its classes with the distribution of the 217 post-’97 landslides. Such a cross-validation is based on the time partitioning of the DSP and is the most natural when time partitioning is available.

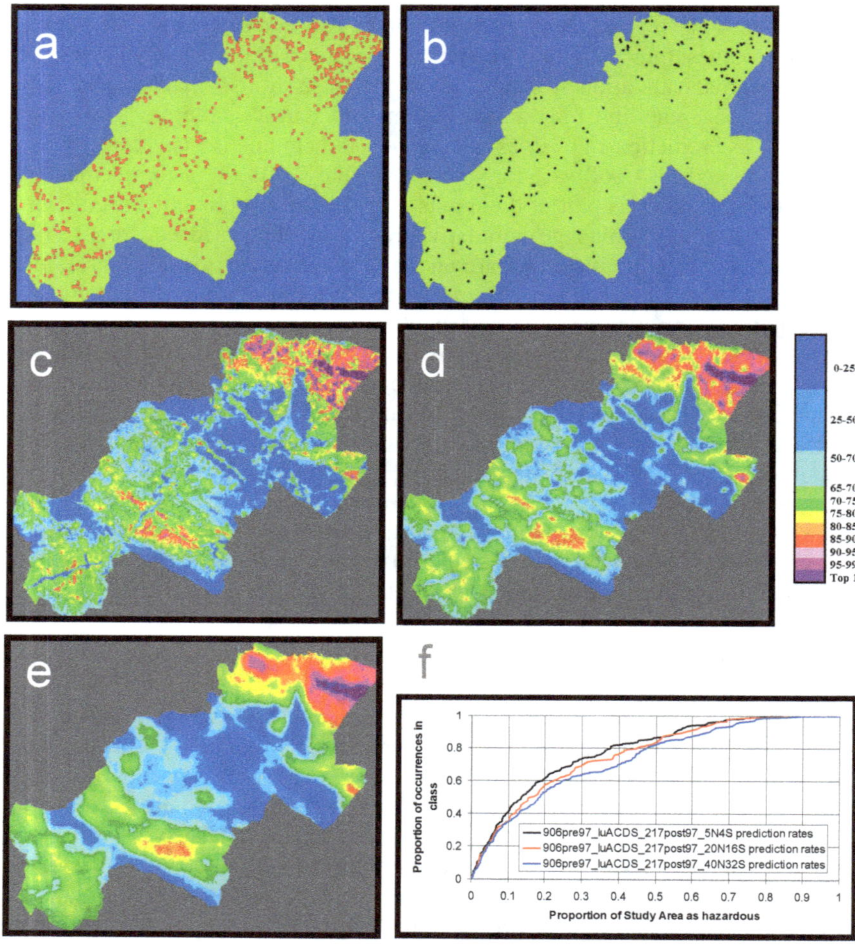

Figure 1: Deba Valley study area. (a) Distribution of the 906-pre'97 shallow translational landslides and associated flows in red and (b) that of the 217-post'97 in black; sizes of trigger zones are exaggerated for visibility. Prediction pattern obtained using the entire distribution of the *1123* landslides as DSP and the *luACDS* as ISPs, with (c) 5*N*4*S*, (d) 20*N*16*S*, and (e) 40*N*32*S*; the legend shows pseudo-colours for different groupings of ranked classes. (f) The three cumulative prediction-rate curves obtained using only the 906-pre'97 landslides as DSP and the *luACDS* as ISPs, to predict the 217-post'97.

Three analyses were performed in which the sets of ISPs were modified as follows:

(i) Using 5 as *number of pixels in the neighbourhood* of categorical boundaries and 4 as *spread parameter* for continuous fields, 5*N*4*S*;

(ii) Using 20 and 16, respectively, 20*N*16*S*; and
(iii) Using 40 and 32, respectively, 40*N*32*S*.

The corresponding ELR prediction patterns are shown in figs 1(c), 1(d) and 1(e), as pseudo-colour images with selected groupings, fixed for visibility, of 200 ranked classes of equal area, each of 0.5% of the study area as default. The loss of detail due to the parameters 5*N*4*S*, 20*N*16*S* and 40*N*32*S* can be seen.

To study the relative quality of those prediction patterns, three more prediction patterns were obtained using only the 906 pre-'97 as DSP and representing the distribution of the remaining 217 post-'97 in them as cumulative prediction-rate curves, as shown in fig. 1(f). On the horizontal axis is the proportion of study area classified as hazardous in decreasing order of 200 ranked values and on the vertical axis the corresponding proportion of the 217 post-'97 landslides. The steeper is the curve at the origin, the better is the prediction pattern: the cumulative curve can be interpreted in terms of cost-benefit. It can be easily seen that the prediction-rate curve with 5*N*4*S* is better than that with 20*N*16*S*, which in turn is better that that with 40*N*32*S*. However, it can be seen that even the prediction-rate curve with 5*N*4*S* indicates a not particularly good prediction. This is due to the variety of geological/land use settings of the study area elongated from the sea coast line to the northeast to inland towards southwest. ELR values are higher for muddy flysch, marly limestone and calcareous flysch lithologies, and lower for other marly units, poorly graded gravel and silty sands, and again higher for grasslands and cultivated areas. All other categorical units and all continuous fields contribute marginally to the ELR values.

In the present analyses, for instance, 0.1 of the study area with the highest predicted values contains 0.42 of the *217*, 0.2 contains 0.61 and 0.3 contains 0.74 with 5*N*4*S*. The corresponding proportions with 20*N*16*S* and 40*N*32*S* are 0.37, 0.58, 0.70, and 0.36, 0.54, 0.60, respectively. The cumulative prediction-rate curves in fig. 1(f) are critical to estimate the probability of occurrence for each class and each pixel for risk assessment later on using an appropriate scenario. This part of the analysis can be considered as preparation of hazard prediction for risk assessment. A complete procedure for it is now considered in eight steps. Table 1 describes the steps of a general framework to be considered for favourability modelling of hazard and risk.

In the procedure described in table 1, the eight steps require decisions, selection of parameters, testing of different alternatives, assumptions and scenarios. For instance, besides the selection of the study area, what becomes critical are those decisions of choosing the DSP, the ISPs, of partitioning the study area and/or the DSP, of selecting a threshold for the uncertainty of class membership and of interpreting the classes of costs in the risk pattern.

Table 1: Eight steps for modelling hazard prediction and risk assessment with favourability modelling, applied to *luACDS* ISPs with parameters 5*N*4*S*, 20*N*16*S* and 40*N*32*S*, using 1123, 906 and 217 landslide distributions as DSP.

Step	Description
1	**Prediction pattern 1**: use distribution of set of **1123** landslides as DSP and the six ISPs, **luACDS,** to be used in all predictions; figs 1(a), 1(b).
2	**Prediction pattern 2**: use the distribution of only the **906** pre-'97 landslides and cross-validate with that of the **217** post-'97 to get the prediction-rate curve. It provides a measure of relative quality; fig. 1(f).
3	**Target and uncertainty patterns**: iterative predictions and cross-validations to establish prediction robustness. Example: exclude successively 50 landslides from the **906** pre-'97 and generate 18 prediction patterns. Using rank-based statistics, the median values of the 18 values per pixel provides the Target and the range of ranks the Uncertainty pattern of class membership; figs 2(a,b), 3(a,b) and 4(a,b).
4	**Combination pattern of uncertainty and target patterns**: selection of reasonable or convenient proportion of lower values of Uncertainty to isolate the corresponding Target values, e.g., 50%; figs 2(c), 3(c) and 4(c).
5	**Introduction of socioeconomic elements at risk**: categorical digital images of identical resolution with a associated tables of values and vulnerabilities, for roads, buildings and land uses. Not shown here.
6	**Modelling the prediction-rate table**: table from **Step 2** is transformed into a table of probability of occurrence for each class and for each pixel, if necessary converted to a monotonically non decreasing function and/or by a fitted exponential function; figs 2(d), 3(d) and 4(d). The example of scenario used assumes that an occurrence rate of **217** one-pixel landslides in 4 years remains constant for the next 50 years since 2001 (217/4 = 54.25×50 = 2712.5 ≈ 2713 one-pixels landslides expected).
7	**Risk pattern**: it is generated using the prediction pattern from **Step 1**, the elements exposed, values and vulnerabilities from **Step 5**, and the modelled probability of occurrence function from **Step 6**. The risk equation is computed and aggregated for roads, building and land uses. In it classes of expected damage per 10 m pixels in € are tentatively reclassified in 10 groups from 100 classes; figs 2(e), 3(e) and 4(e). Only the highest 4 1% classes have values > 1 € and are indicated by black pixels.
8	**Combination pattern of uncertainty and risk patterns**: use the Combination pattern of 50% lower values of uncertainty in **Step 4** to obtain the corresponding Combination uncertainty/risk pattern with values corresponding to relatively low uncertainty; figs 2(f), 3(f) and 4(f).

The risk patterns in the three experiments are visibly different, however, in all of them it is the highest 4 classes that represent risk values greater than 1.00 €/pixel. Table 2 shows the respective expected cost of a pixel for those classes. The expected total costs for the Risk patterns in the study area are as follows: 9,104,146 € (5*N*4*S*), 9,649,109 (20*N*16*S*) and 12,816,878 (40*N*32*S*).

Table 2: Expected cost of a pixel > 1 € for the four highest risk classes in experiments 5*N*4*S*, 20*N*16*S* and 40*N*32*S*.

Class	**€**	**5*N*4*S* €**	**20*N*16*S* €**	**40*N*32*S* €**
97	> 1	1.64	1.08	1.50
98	> 2	2.94	2.33	2.68
99	> 5	5.29	5.23	5.04
100	> 10	11.98	16.39	12.33

5 Concluding remarks

Many aspects of spatial prediction modelling provide challenges that demand experimentations on representative databases. Generalizing the procedure applied, some or all of the following points can be questioned:

(i) *Database*: Study area, direct and indirect supporting patterns, homogeneity of occurrences and settings, time/space partitions for cross-validation.
(ii) *Prediction patterns*: How good are they? Were they over-fitted? Is it enough to use prediction-rate tables via relative ranking?
(iii) *Modelling*: What happens if conditional independence assumptions, required by most models, are not respected?
(iv) *Scenarios*: How to formulate and construct them to estimate the probability of occurrence essential for risk assessment? How far back in time should one look at the statistics of occurrences and to predict how far in the future?
(v) *Expected costs*: How representative are such costs? How to classify and interpret them, and display their spatial configuration?

Because of these many challenging aspects due to variability of all points discussed, empirical analyses of databases more in demand were considered than the development of new mathematical models. The distribution and sharing of spatial databases among expert organizations or researchers is perhaps a more promising way to deepen insight on how to manage spatial data for modelling the future as to the discovery of new natural resources or to the identification of future hazardous zones. Such a sharing is on its own a considerable challenge.

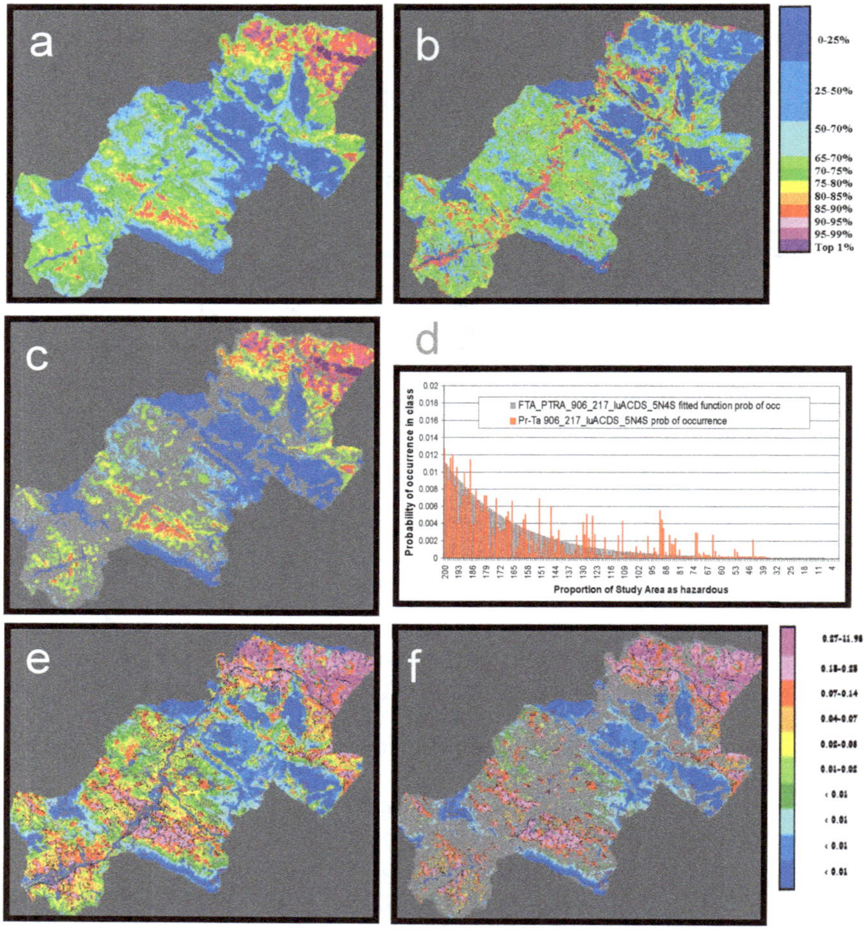

Figure 2: Target, uncertainty and risk patterns with probability of occurrence histograms for the Deba Valley study area for analyses with *5N4S* parameters for the *luACDS* ISPs. (a) Target pattern obtained using the distribution of the 906-pre'97 landslides and the 906-50×18 iterative cross-validation, (b) Uncertainty pattern and (c) 50% Uncertainty/Target Combination pattern. (d) Transformation of the corresponding prediction-rate in fig. 1(d), into a probability of occurrence histogram in red, and in gray the fitted function to its monotonically decreasing transformation. (e) Risk pattern with the four classes with estimated costs > 1 € in black, and (f) the corresponding 50% Uncertainty/Risk Combination pattern of (c). Ten classes are in the risk legend with expected costs in €.

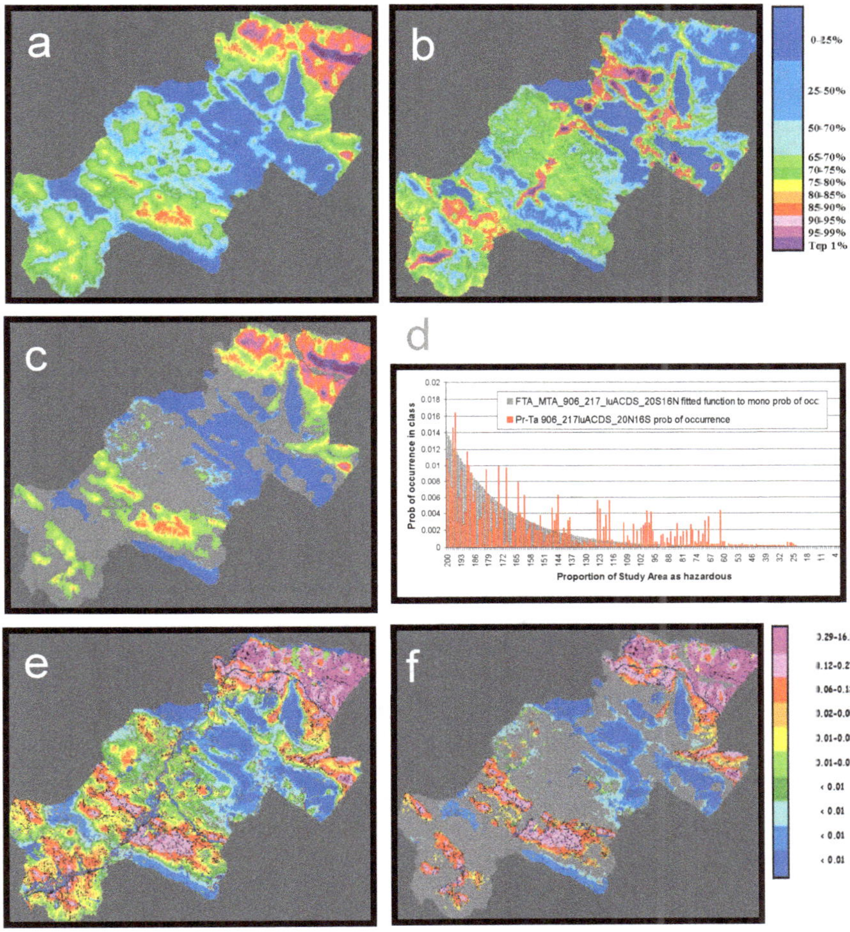

Figure 3: Target, uncertainty and risk patterns with probability of occurrence histograms for the Deba Valley study area for analyses with 20*N*16*S* parameters for the *luACDS* ISPs. (a) Target pattern obtained using the distribution of the 906-pre'97 landslides and the 906-50×18 iterative cross-validation, (b) Uncertainty pattern and (c) 50% Uncertainty/Target Combination pattern. (d) Transformation of the corresponding prediction-rate in fig. 1(d) into a probability of occurrence histogram in red and in gray the fitted function to its monotonically decreasing transformation. (e) Risk pattern with the four classes with estimated costs > 1 € in black, and (f) the corresponding 50% Uncertainty/Risk Combination pattern of (c). Ten classes are in the risk legend with expected costs in €.

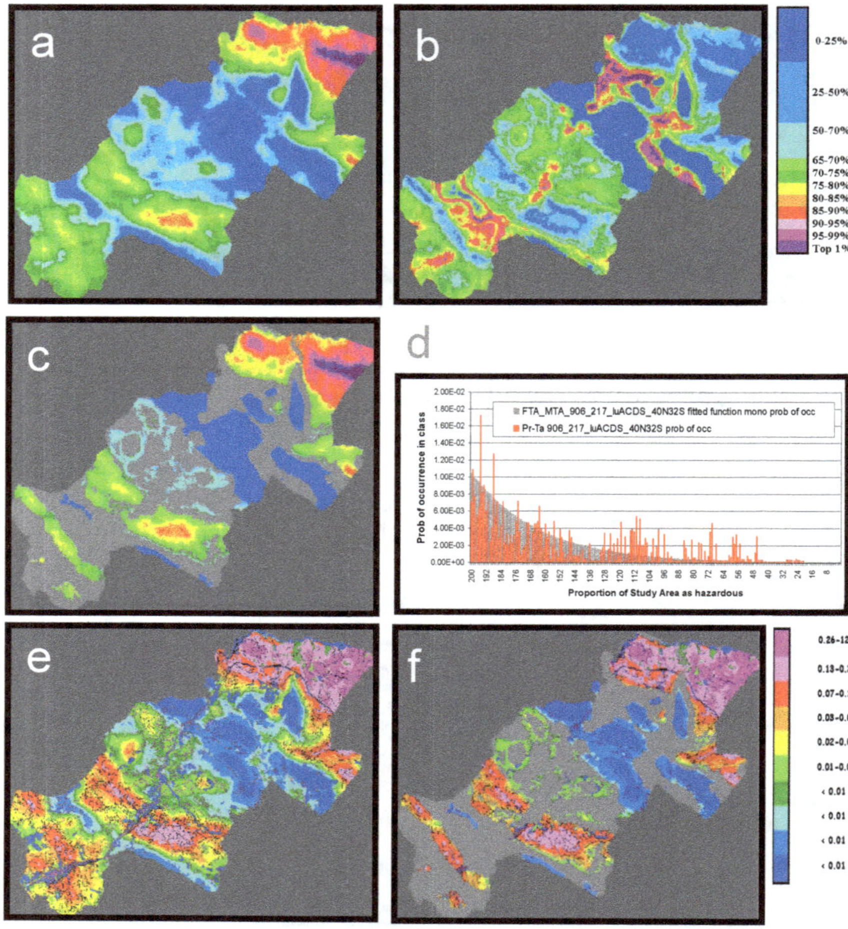

Figure 4: Target, uncertainty and risk patterns with probability of occurrence histograms for the Deba Valley study area for analyses with 40*N*32*S* parameters for the *luACDS* ISPs. (a) Target pattern obtained using the distribution of the 906-pre'97 landslides and the 906-50×18 iterative cross-validation, (b) Uncertainty pattern and (c) 50% Uncertainty/Target Combination pattern. (d) Transformation of the corresponding prediction-rate in fig. 1(d) into a probability of occurrence histogram in red and in gray the fitted function to its monotonically decreasing transformation. (e) Risk pattern with the four classes with estimated costs > 1 € in black, and (f) the corresponding 50% Uncertainty/Risk Combination pattern of (c). Ten classes are in the risk legend with expected costs in €.

References

[1] Remondo, J., González-Díez, A., Díaz de Terán, J.R. & Cendrero, A., Landslide susceptibility models utilising spatial data analysis techniques. A case study from the lower Deba valley, Guipúzcoa (Spain). *Natural Hazards*, **30**, pp. 267-279, 2003.

[2] Remondo, J., González-Díez, A., Díaz de Terán, J.R., Cendrero, A., Fabbri, A. & Chung, C.F., Validation of landslide susceptibility maps; examples and applications from a case study in northern Spain. *Natural Hazards*, **30**, pp. 437-449, 2003.

[3] Remondo, J., Bonachea, J. & Cendrero, A., A statistical approach to landslide risk modelling at basin scale: from landslide susceptibility to quantitative risk assessment. *Landslides*, **2(4)**, pp. 321-328, 2005.

[4] Fabbri, A.G. & Chung, C.-J., Training decision-makers in hazard spatial prediction and risk assessment: ideas, tools, strategies and challenges. *Disaster Management and Human Health Risk*, eds. K. Duncan & C.A. Brebbia, WIT Press: Southampton, pp. 285-296, 2009.

[5] Chung, C.F., Using likelihood ratio functions for modelling the conditional probability of occurrence of future landslides for risk assessment. *Computers and Geosciences*, **32**, pp. 1052-1065, 2006.

[6] Fabbri, A.G. & Chung, C.-J., *On blind tests and spatial prediction models. Natural Resources Research*, **17(2)**, pp. 107-118, 2008.

[7] http://www.spatialmodels.com

Author index

WITPRESS ...for scientists by scientists

Disaster Management and Human Health Risk IV

Reducing Risk, Improving Outcomes

Edited by: ***Ş.M. SENER****, Istanbul Technical University, Turkey;* ***C.A. BREBBIA****, Wessex Institute of Technology, UK and* ***Ö. ÖZÇEVIK****, Istanbul Technical University, Turkey*

As human population has continued to concentrate in urban areas, the number of people and the value of property affected by disasters, both natural and human-generated, have grown as well. Earthquakes, floods, hurricanes, cyclones, tornadoes, and forest fires have all taken their toll. So have such anthropogenic disasters as pipeline failures, industrial spills, and terrorist attacks.

The contents of this volume consist of papers presented at the fourth in a series of conferences convened to assess the potential risk from various disasters and discuss ways to prevent or mitigate damage. The papers have been contributed by experts on public health, security, and disaster management from academia, industry, and government.

Topics covered include: Disaster Analysis; Disaster Monitoring and Mitigation; Emergency Preparedness; Risk Assessment and Mitigation; Risk and Security; Safety and Resilience; Socio-economic Issues; Case Studies.

WIT Transactions on The Built Environment, Vol 150
ISBN: 978-1-84564-926-5 **eISBN: 978-1-84564-927-2**
Published 2015 / 360pp

WIT eLibrary

Home of the Transactions of the Wessex Institute, the WIT electronic-library provides the international scientific community with immediate and permanent access to individual papers presented at WIT conferences. Visitors to the WIT eLibrary can freely browse and search abstracts of all papers in the collection before progressing to download their full text.

Visit the WIT eLibrary at
http://library.witpress.com

www.ingramcontent.com/pod-product-compliance
Lightning Source LLC
LaVergne TN
LVHW011849200926
848484LV00002B/425
* 9 7 8 1 7 8 4 6 6 2 2 9 5 *